# Carbohydrate
# analysis

# TITLES PUBLISHED IN
## —— THE ——
## PRACTICAL APPROACH
## —— SERIES ——

# Carbohydrate analysis

## a practical approach

Edited by

## M F Chaplin

Department of Biotechnology, South Bank Polytechnic,
London SE1 0AA, UK

## J F Kennedy

Research Laboratory for the Chemistry of Bioactive
Carbohydrates and Proteins, Department of Chemistry,
University of Birmingham, Birmingham B15 2TT, UK

 **IRL PRESS**
Oxford · Washington DC

IRL Press Limited,
P.O. Box 1,
Eynsham,
Oxford OX8 1JJ,
England

British Library Cataloguing in Publication Data

Carbohydrate analysis: a practical approach.
—(The Practical approach series)
  1.  Food—Analysis   2.   Carbohydrates—Analysis
  I.   Chaplin,M.F.   II.   Kennedy,John F. (John Frederick)   III.   Series
  641.1 3    TX553.C28

ISBN 0-947946-44-6 (softbound)
ISBN 0-947946-68-3 (hardbound)

Printed in England by Information Printing, Oxford.

# Preface

The involvement of carbohydrates in biological processes has greatly fuelled the current interest in this diverse range of molecules. This has resulted in a vast literature covering numerous analytical methods for these carbohydrates. The size of the literature arises in part from the large number of classes of carbohydrates — macromolecular and monomolecular — basic, neutral and acidic — derivatized and underivatized — and the wide range of presentation of carbohydrates in everyday life — biosynthesis or chemicosynthesis — structural or gelatinous — edible or unmetabolizable. Superimpose upon just these selected aspects the notion of primary, secondary, tertiary and quaternary structure and the fact that the number of ways of covalently joining two carbohydrate monomers is at least one order of magnitude greater than the similar joining of two amino acids, then it is clear that analysis of the field of carbohydrates is, and must be, vast. For this reason, experts in particular areas of the field have been asked to contribute to this book — without them it would not have been possible.

It is not always obvious to the researcher or analyser which method, or even physical technique, is most appropriate to a particular investigation. This book has been produced in order to answer the need for a handbook of laboratory protocols in this field. It gives details of the approach needed to analyse a wide variety of carbohydrates and carbohydrate-containing molecules. We have attempted to show how particular analytical problems should be tackled, describing the most suitable, well tried and trusted methods in exact practical detail.

Chapters are arranged on the basis of carbohydrate moiety to facilitate choice of analytical method for specific applications. Thus the chapters act as a ready source of reference at the initial planning stage for the choice of approach to situations where carbohydrate analysis is required.

We are indebted to our co-authors, who have made this book a useful source of reference for carbohydrate analytical protocols which should find active and practical use within the laboratory.

Martin Chaplin and John F.Kennedy

# Contributors

S.Bouquelet
*Université des Sciences et Techniques de Lille I, Laboratoire de Chimie Biologique, 59655 Villeneuve D'Ascq Cedex, France*

S.L.Carney
*The Mathilda and Terence Kennedy Institute of Rheumatology, 6 Bute Gardens, Hammersmith, London W6 7DW, UK*

M.F.Chaplin
*Department of Biotechnology, South Bank Polytechnic, London SE1 0AA, UK*

H.Debray
*Université des Sciences et Techniques de Lille I, Laboratoire de Chimie Biologique, 59655 Villeneuve D'Ascq Cedex, France*

B.Fournet
*Université des Sciences et Techniques de Lille I, Laboratoire de Chimie Biologique, 59655 Villeneuve D'Ascq Cedex, France*

J.F.Kennedy
*Department of Chemistry, University of Birmingham, Birmingham B15 2TT, UK*

J.Montreuil
*Université des Sciences et Techniques de Lille I, Laboratoire de Chimie Biologique, 59655 Villeneuve D'Ascq Cedex, France*

I.M.Morrison
*Hannah Research Institute, Ayr KA6 5HL, UK*

J.H.Pazur
*Department of Biochemistry, Pennsylvania State University, Paul M Althouse Laboratory, University Park, PA 16802, USA*

G.Spik
*Université des Sciences et Techniques de Lille I, Laboratoire de Chimie Biologique, 59655 Villenuve D'Ascq Cedex, France*

G.Strecker
*Université des Sciences et Techniques de Lille I, Laboratoire de Chimie Biologique, 59655 Villeneuve D'Ascq Cedex, France*

C.A.White
*Chembiotech Ltd, Institute of Research and Development, Vincent Drive, Birmingham B15 2SQ, UK*

# Contents

# Abbreviations

| | |
|---|---|
| c.d. | circular dichroism |
| CIAE | crossed immuno-affino electrophoresis |
| DEAE | diethylaminoethyl |
| DMSO | dimethylsulphoxide |
| d.p. | degree of polymerization |
| DPO | diphenyloxazole |
| EDTA | ethylenediaminetetraacetic acid |
| e.i.-m.s. | electron impact mass spectrometry |
| f.a.b.-m.s. | fast atom bombardment mass spectrometry |
| GAG | glycosaminoglycan |
| g.l.c. | gas-liquid chromatography |
| g.p.c. | gel permeation chromatography |
| HA | hyaluronic acid |
| HABR | hyaluronic acid binding region |
| h.p.l.c. | high performance liquid chromatography |
| i.r. | infra-red spectroscopy |
| LP | link protein |
| m.s. | mass spectroscopy |
| NEM | $N$-ethylmaleimide |
| n.m.r. | nuclear magnetic resonance |
| PBS | phosphate-buffered saline |
| p.c. | paper chromatography |
| PMSF | phenylmethylsulphonyl fluoride |
| SDS | sodium dodecylsulphate |
| SDS-PAGE | polyacrylamide gel electrophoresis in the presence of SDS |
| TEMED | $N,N,N',N'$-tetramethylethylenediamine |
| TFA | trifluoroacetic acid and trifluoroacetates |
| t.l.c. | thin-layer chromatography |
| WCOT | wall coated open tubular |

# CHAPTER 1

# Monosaccharides

MARTIN F.CHAPLIN

## 1. INTRODUCTION

Monosaccharides exhibit a great variety of structural types with vastly different chemical and physical properties. In contrast to the several methods available for the analysis of the other major class of biological molecules, the amino acids, there is no single method which is suitable for the quantitative or qualitative analysis of all monosaccharides. The method of choice will depend on a number of factors including the accuracy required and resources available. There are two techniques available for the quantitative analysis of mixtures of monosaccharides, high performance liquid chromatography (h.p.l.c.) and gas-liquid chromatography (g.l.c.). If only a qualitative analysis is needed, either paper chromatography (p.c.) or thin-layer chromatography (t.l.c.) may be used. Single monosaccharides may be identified by means of mass spectrometry (m.s.), infra-red spectroscopy (i.r.) or proton or carbon-13 nuclear magnetic resonance ($^1$H-n.m.r., $^{13}$C-n.m.r.), and quantitated by use of specific colorimetric or enzymatic assays. An extensive and growing literature exists which describes methods within these techniques. In this chapter a number of these methods have been chosen and described. They encompass the most commonly occurring analytical circumstances but several volumes would be needed if a fully comprehensive text was to be presented.

## 2. COLORIMETRIC ASSAY

A number of colorimetric assays are presented for the main classes of monosaccharides, to be followed by an important specific enzymatic assay. A standard format has been chosen to describe these assays. The sensitivity describes the range over which the assay is fairly linear with a maximum absorbance of 1.0 OD unit for a 1 cm path length cuvette. Distilled or deionized water is used throughout for aqueous solutions. The final volume, for spectrophotometric measurement, has been kept within the $1-2$ ml range to allow microanalytical cuvettes to be used, but larger volumes can be arranged by increasing the sample and reagent volumes in proportion. In general the protocols should be followed in as reproducible a manner as possible.

### 2.1 Hexose

2.1.1 *L-Cysteine sulphuric acid assay* (1)

*Sensitivity:* $\sim 0.2-20$ μg glucose in 200 μl ($\sim 5-500$ μM).

*Final volume*: 1.2 ml.

*Reagents:*

(A)  Ice-cold 86% v/v sulphuric acid, prepared by the very careful addition of 860 ml good quality concentrated sulphuric acid to 140 ml of water. This addition generates considerable heat and should be carried out within a fume cupboard (using protective glasses and clothing) at least the day before the solution is to be used. The reagent is stable if kept dust-free in a stoppered containers.

(B)  Freshly prepared solution of L-cysteine hydrochloride (700 mg/l) in reagent A.

*Method*

(i)  To samples, standards and control solutions (200 $\mu$l) containing up to 20 $\mu$g neutral carbohydrate cautiously and reproducibly add 1.0 ml of reagent B, with immediate thorough mixing in an ice bath.

(ii)  Heat at 100°C for 3 min in a glass stoppered test tube.

(iii)  Rapidly cool the mixture to room temperature.

(iv)  Determine the absorbance at 415 nm.

*Comments.* Assay tubes must be scrupulously clean for this assay as it measures both monomeric and polymeric neutral carbohydrate and is, therefore, interfered with by extraneous glucose in airborne dust. Deoxy sugars, heptoses, pentoses and uronic acids all produce colour in this assay with absorbance maxima in the range 380−430 nm. The stability of this colour varies between the carbohydrates and distinctive shifts in absorbance may be obtained on standing for 24−48 h. Azide, heavy metal ions and non-aldose reducing substances interfere in this assay. Mixtures of carbohydrates giving different absorbance maxima can be analysed by use of two standard determinations as outlined in Section 2.9.

### 2.1.2 Phenol-sulphuric acid assay (2)

*Sensitivity:* ~1−60 $\mu$g glucose in 200 $\mu$l (~30 $\mu$M−2 mM).

*Final volume*: 1.4 ml.

*Reagents:*

(A)  Phenol dissolved in water (5% w/v). This solution is stable indefinitely.

(B)  Concentrated sulphuric acid.

*Method*

(i)  Mix samples, standards and control solutions (200 $\mu$l containing up to 100 $\mu$g carbohydrate) with 200 $\mu$l of reagent A.

(ii)  Add 1.0 ml of reagent B rapidly and directly to the solution surface without touching the sides of the tube.

(iii)  Leave the solutions undisturbed for 10 min before shaking vigorously.

(iv)  Determine the absorbances at 490 nm after a further 30 min.

*Comments.* Other aldoses, ketoses and alduronic acids respond to different degrees. Protein, cysteine, non-carbohydrate reducing agents, heavy metal ions and azide interfere with this assay. However, it remains useful as a rapid non specific method for the detection of neutral carbohydrate in column eluates.

## 2.2 **Reducing sugar**

### 2.2.1 *Neocuproine assay* (3)

*Sensitivity:* ~20 ng−2 μg glucose in 200 μl (~0.5−50 μM).

*Final volume*: 2.0 ml.

*Reagents:*

(A)    Dissolve 40 g of anhydrous sodium carbonate, 16 g of glycine and 450 mg cupric sulphate pentahydrate in 600 ml water and make up to 1 litre with more water. The solution is stable indefinitely.

(B)    Dissolve 0.15 g neocuproine hydrochloride in 100 ml water. This solution is stable for months if stored in a dark bottle.

*Method*

(i)     To samples, standards and controls, (200 μl) add 400 μl reagent A and 400 μl reagent B. Mix well.

(ii)    Heat the solutions at 100°C for exactly 12 min and cool rapidly to room temperature.

(iii)   Add 1.0 ml of water, washing down any condensate formed.

(iv)   Mix well and determine the absorbance at 450 nm.

*Comments.* If a precipitate forms during the final cooling it may simply be redissolved by slight warming. Non-carbohydrate reducing agents interfere in this assay.

### 2.2.2 *Dinitrosalicylic acid assay* (4,5)

*Sensitivity:* ~5−500 μg glucose in 100 μl (~0.3−30 mM).

*Final volume*: 1.1 ml.

*Reagent:* Dissolve 0.25 g of 3,5-dinitrosalicylic acid and 75 g sodium potassium tartrate (Rochelle salt) in 50 ml 2 M sodium hydroxide (made by dissolving 4 g NaOH in 50 ml water) and dilute to 250 ml with water. This is stable for several weeks.

*Method*

(i)     To samples, standards and controls (100 μl) add 1.0 ml of the reagent. Mix well.

(ii)    Heat the mixtures at 100°C for 10 min.

(iii)   After rapid cooling to room temperature, determine the absorbance at 570 nm.

*Comments.* Dissolved molecular oxygen interferes with this assay. This may be overcome either by purging the assay solutions with nitrogen or helium prior to the assay or by the addition of a fixed known small amount of glucose (~20 μg) to all samples, in order to raise the total reducing sugars concentration above a critically low value.

Non-carbohydrate reducing agents also interfere with this assay. Some metal ions [e.g. manganous, cobalt (II), and calcium] may increase the assay response.

## 2.3 **Pentose**

### 2.3.1 *Ferric-orcinol assay* (4)

*Sensitivity:* ~200 ng−20 μg xylose in 200 μl (~7−700 μM).

*Final volume:* 1.6 ml.

*Reagents:*

(A)   Trichloroacetic acid solution in water (10% w/v). This is stable indefinitely.
(B)   Freshly prepared solution of ferric ammonium sulphate (1.15% w/v) and orcinol (0.2% w/v) in 9.6 M hydrochloric acid (made by diluting five parts concentrated HCl with one part water).

*Method*

(i)    Mix the samples, standards and control solutions (200 $\mu$l) containing up to 40 $\mu$g pentose with 200 $\mu$l of reagent A.
(ii)   Heat at 100°C for 15 min.
(iii)  Cool the solution rapidly to room temperature. Add reagent B (1.2 ml) and mix well.
(iv)   Reheat the solution at 100°C for a further 20 min.
(v)    Cool the solution to room temperature and determine the absorbance at 660 nm.

*Comments.* Hexoses interfere in this assay but can be accounted for by additionally determining the absorbance at 520 nm at which wavelength they have a strong absorbance. It is recommended that both hexose and pentose standards are used where the hexose content of the samples might be considerable.

## 2.4 Ketose

2.4.1 *Phenol-boric acid-sulphuric acid assay* (6)

*Sensitivity:* ~0.1−9 $\mu$g fructose in 100 $\mu$l (~30−500 $\mu$M).

*Final volume:* 2.0 ml.

*Reagents:*

(A)   Dissolve 2.5 g phenol (recrystallized from methanol and ethanol) in 50 ml water. Add 1.0 ml of acetone dropwise with constant stirring over a period of 10 min and stir the mixture for a further 10 min at room temperature. Dissolve 2.0 g of boric acid in the mixture. The reagent is stable for at least 2 weeks at 4°C.
(B)   Concentrated sulphuric acid.

*Method*

(i)    Mix the samples, standards and controls (100 $\mu$l) with 0.5 ml of reagent A and then rapidly add 1.4 ml of reagent B directly to the surface, avoiding the sides of the tubes.
(ii)   After thorough mixing, leave the solutions for 5 min at room temperature.
(iii)  Incubate at 37°C for 1 h.
(iv)   Determine the absorbance at 568 nm.

*Comments.* Different ketoses give differing absorbances in this assay. Interferences from non-ketose carbohydrate is slight (<1%) to non-existent. The reproducibility of this assay is strongly dependent on the manner of the addition of the sulphuric acid.

## 2.5 Hexosamine

2.5.1 *Morgan-Elson assay* (7)

*Sensitivity:* ~60 ng−6 $\mu$g 2-acetamido-2-deoxy-D-glucose in 250 $\mu$l (~1−110 $\mu$M).

*Final volume:* 1.8 ml.

*Reagents:*

(A)  Dissolve 6.1 g of di-potassium tetraborate tetrahydrate in 80 ml of water and make up to 100 ml with water.

(B)  Add 1.5 ml of water to 11 ml of concentrated hydrochloric acid. Add a further 87.5 ml of glacial acetic acid and dissolve 10 g of 4-(*N,N*-dimethylamino)-benzaldehyde in this mixture. This solution may be stored for several weeks. Dilute 10 ml to 100 ml with galacial acetic acid immediately prior to use.

*Method*

(i)    Add samples, standards and controls (250 μl) to 50 μl of reagent A.
(ii)   Heat each mixture at 100°C for 3 min.
(iii)  After cooling rapidly to room temperature, add 1.5 ml of reagent B, washing down any condensate formed.
(iv)   Incubate the samples at 37°C for 20 min.
(v)    After cooling to room temperature, determine the absorbances at 585 nm.

*Comments.* 2-Acetamido-2-deoxy-D-galactose gives only one third the response of 2-acetamido-2-deoxy-D-glucose in this assay. Free amino-hexoses may be *N*-acetylated prior to this assay by the addition of one part of freshly prepared 1.5% (v/v) acetic anhydride in acetone to eight parts of the aqueous solution and leaving for 5 min at room temperature.

## 2.6 Uronic acid

### 2.6.1 *Carbazole assay* (8)

*Sensitivity:* ~200 ng−20 μg D-glucurono-6,3-lactone in 250 μl (~4−400 μM).

*Final volume:* 1.8 ml.

*Reagents:*

(A)  Dissolve 0.95 g of sodium tetraborate decahydrate in 2.0 ml of hot water and add 98 ml of ice-cold concentrated sulphuric acid carefully with stirring. This reagent is stable indefinitely if refrigerated.

(B)  Dissolve 125 mg of carbazole (recrystallized from ethanol) in 100 ml of absolute ethanol to give a stable reagent.

*Method*

(i)     Cool the samples, standards and controls (250 μl) in an ice bath.
(ii)    Add ice-cold reagent A (1.5 ml) with mixing and cooling in the ice bath.
(iii)   Heat the mixtures at 100°C for 10 min.
(iv)    Cool rapidly in the ice-bath.
(v)     Add 50 μl of reagent B and mix well.
(vi)    Reheat at 100°C for 15 min.
(vii)   Cool rapidly to room temperature and determine the absorbance at 525 nm.

*Comments.* Neutral carbohydrates interfere with this assay to a greater (~10% on a molar basis for hexoses) or lesser extent (~1% on a molar basis for pentoses). However both types of interference can be eliminated by use of appropriate controls as they give significantly different absorption spectra. Cysteine and other thiols increase the response of the assay but large amounts of protein may depress the colour development. Different uronic acids give different responses in this assay.

## 2.7 Sialic acid

### 2.7.1 Warren assay (9)

*Sensitivity:* 80 ng$-$8 $\mu$g *N*-acetyl neuraminic acid in 80 $\mu$l (3$-$300 $\mu$M).

*Final volume:* 1.0 ml.

*Reagents:*

(A)   Dissolve 4.278 g of sodium metaperiodate in 4.0 ml water. Add 58 ml of concentrated orthophosphoric acid and make up to 100 ml with water. This reagent is stable indefinitely.

(B)   Dissolve 10 g of sodium arsenite and 7.1 g of sodium sulphate in 0.1 M sulphuric acid (made by carefully diluting 5.7 ml concentrated sulphuric acid to 1 litre with water) to a total volume of 100 ml. The solution is stable indefinitely.

(C)   Dissolve 1.2 g of 2-thiobarbituric acid and 14.2 g of sodium sulphate in water to a total volume of 200 ml. This reagent is stable for several weeks but eventually forms a yellow precipitate which indicates the need for its renewal.

(D)   Redistilled cyclohexanone. This is stable for several months or until noticeably discoloured.

*Method*

(i)      To the samples, standards and controls (80 $\mu$l) add 40 $\mu$l reagent A and mix well.

(ii)     Leave at room temperature for 20 min.

(iii)    Add 400 $\mu$l reagent B and then shake the tubes vigorously to expel the yellow-coloured iodine.

(iv)    Leave for a further 5 min at room temperature.

(v)     Add 1.2 ml reagent C, shake the tubes, stopper them and heat at 100°C for 15 min.

(vi)    Cool rapidly to room temperature.

(vii)   Extract the chromophore into 1.0 ml of reagent D by vigorous shaking. Centrifuge the solutions using a bench centrifuge for a few minutes in order to properly separate the two layers.

(viii)  Determine the absorbance of the upper cyclohexanone layer at 549 nm.

*Comments.* DNA, 2-deoxy-D-ribose and substances producing malondialdehyde on periodate oxidation interfere in this assay. This may be circumvented by additionally determining the absorbance at 532 nm and calculating from the resultant data (see Section 2.9). L-Fucose reduces the expected absorbance of this assay. Methoxyneuraminic acid and some acetylated neuraminic acids give no colour in this assay and the resorcinol-HCl assay (10) should be used if the presence of any of these is suspected. The assay is less specific than the Warren assay, however, and is not recommended for general use.

## 2.8 Enzymatic methods

### 2.8.1 Hexokinase/dehydrogeanse assay (11)

*Sensitivity:* 400 ng$-$40 $\mu$g glucose in 100 $\mu$l (20 $\mu$M$-$2 mM).

*Final volume:* 1.4 ml.

*Reagents:*

(A)    0.33 M triethanolamine, 4.3 mM $Mg^{2+}$. Dissolve 6.0 g triethanolamine hydrochloride and 0.11 g $MgSO_4.7H_2O$ in 80 ml water. Adjust the pH to 7.6 with concentrated NaOH solution ($\sim 20\%$ w/v) and make up the solution to 100 ml with water. This buffer is stable for a month at 4°C.

(B)    5.5 mM NADP. Dissolve NADP (disodium salt, 4.3 mg/ml) in distilled water. This solution is stable for a month at 4°C.

(C)    35 mM ATP, 0.26 M $NaHCO_3$. Dissolve ATP (disodium salt hydrate, 22 mg/ml) and $NaHCO_3$ (22 mg/ml) in water. This solution is stable for a month at 4°C.

(D)    3.2 M ammonium sulphate. Add 0.6 g ammonium sulphate to 1 ml of water and allow to dissolve. This solution is stable.

(E)    Dissolve hexokinase (ATP: D-hexose-6-phosphotransferase, EC 2.7.1.1 ex. yeast, 280 iu/ml, 2 mg/ml approx.) and glucose-6-phosphate dehydrogenase (D-glucose-6-phosphate: NADP 1-oxidoreductase, EC 1.1.1.49, ex. yeast, 140 iu/ml, 1 mg/ml approx.) in solution D. This solution is stable for a month at 4°C.

Some of these reagents are available in a kit form from manufacturers.

*Method*

(i)    Add each sample and standard solution (100 μl) to a mixture containing 1.0 ml buffer solution A, 100 μl reagent B and 100 μl reagent C. Mix well.

(ii)   Start the reaction of 100 μl of enzyme solution E. The control solutions lack enzyme and so should consist of sample solution (100 μl) plus reagents A (1.1 ml), B (100 μl) and C (100 μl).

(iii)  After further mixing, incubate the solutions at 37°C for 30 min.

(iv)   Cool and determine the absorbance at 340 nm. The reaction should have stopped at this stage. However, check whether there is a significant change ($>5\%$) in absorbance at 340 nm after a further 30 min incubation at 37°C. If so, check the reagents and/or extend the incubation time.

*Comments.* The assay can be adapted for the determination of fructose and/or mannose in the presence or absence of glucose. Determine fructose by the addition of 100 μl phosphoglucose isomerase (D-glucose-6-phosphate ketol-isomerase, EC 5.3.1.9, ex. yeast, 65 iu/ml, 0.2 mg/ml approx. in buffer solution A) to the reaction mixture after the glucose content has been determined. Mix this solution, incubate at 37°C for 30 min, cool and re-measure the absorbance at 340 nm. Determine mannose in a similar manner subsequent to the addition of 100 μl phosphomannose isomerase (D-mannose-6-phosphate ketol-isomerase EC 5.3.1.8, ex. yeast, 60 iu/ml, 1 mg/ml approx. in buffer solution A). The sensitivities for fructose and mannose are similar to those for glucose. Starch and sucrose may also be determined after enzymatic conversion to monosaccharides by glucoamylase (12) or invertase (13) respectively. As the enzymes are optimally active at substantially lower pH than hexokinase it is recommended that a separate procedure be adopted whereby the hydrolysed and unhydrolysed samples are analysed in the above assay. Similar assay systems may be set up for the determination of other carbohydrates where the appropriate enzyme is available, for example L-fucose using fucose dehydrogenase (14) and D-galactose using $\beta$-D-galactose dehydrogenase (15). In these cases the buffer solution A should be replaced by a buffer appropriate to the determination and reagents B, C, D and E will all probably be different. The principal of the assay however will remain as the change in absorbance at 340 nm due to the formation of the reduced dinucleotide.

The sensitivity of this spectrophotometric assay may be increased 10- to 100-fold (0.2 – 20 $\mu$M) by using the fluorescence change rather than the absorption change. The excitation wavelength is 340 nm and the fluorescence is emitted at about 465 nm. This improvement in sensitivity is achieved at an extra cost in the care needed for the assay. All solutions should be dust-free and all glassware scrupulously cleaned. The cuvettes should be of low fluorescence glass or quartz and temperature-equilibrated before the determinations are made. Because of the higher background variability, this method is best chosen only when the additional sensitivity over the spectrophotometric assay is essential.

If glucose alone needs to be assayed, the use of a one-enzyme assay system based on glucose dehydrogenase may be used. This method is simpler and cheaper but is, as yet, not as well standardized (16).

## 2.9 Example calculations

A mixture of two components (A, B) can simply be determined spectrophotometrically if each component absorbs maximally at different wavelengths (X nm and Y nm).

(i)  Using a suitable range of standard concentrations for both components separately determine the absorbance produced by each at both wavelengths (i.e. component A at concentration $C_A$ gives absorbance $\Delta A_x$ at wavelength X nm and $\Delta A_y$ at Y nm. Similarly component B at concentration $C_B$ gives absorbance $\Delta B_x$ at X nm and $\Delta B_y$ at Y nm).

(ii)  Determine the absorbance difference between blank and unknown sample S at the same wavelengths X ($\Delta S_X$) and Y ($\Delta S_Y$).

(iii)  The concentration of A in the sample is

$$C_A \times \left( \frac{\Delta S_X \times \Delta B_Y - \Delta S_Y \times \Delta B_X}{\Delta A_X \times \Delta B_Y - \Delta A_Y \times \Delta B_X} \right)$$

The concentration of B in the sample is

$$C_B \times \left( \frac{\Delta S_X \times \Delta A_Y - \Delta S_Y \times \Delta A_X}{\Delta B_X \times \Delta A_Y - \Delta B_Y \times \Delta A_X} \right)$$

For example, for the pentose assay (Section 2.3.1) the absorbance maxima X and Y are 660 nm and 520 nm, respectively. The absorbance of 20 $\mu$g xylose (A) at these maxima are $\Delta A_X = 1.0$ and $\Delta A_Y = 0.25$, whereas 180 $\mu$g of glucose (B) gives $\Delta B_X = 0.25$ and $\Delta B_Y = 0.60$.

## 3. THIN-LAYER CHROMATOGRAPHY (17 – 19)

### 3.1. Experimental approach

Thin-layer chromatography is a simple and rapid technique that is very useful for the preliminary examination of carbohydrate mixtures. A number of solid supports have been used with a vast number of solvents and detection methods for separating a wide range of carbohydrates. Adding to the difficulty of choosing a system for recommendation is the fact that workers in this field never seem to choose the same range of standard sugars to calibrate and compare their systems. It is clear, however, that no single system is available that will separate all possible combinations of carbohydrates. The best plan is to try likely systems until a suitable one is found.

There are two solid supports which have proved themselves to be particularly useful, microcrystalline cellulose and silica gel. Cellulose separates essentially by liquid-liquid partition. The sugar is distributed between the mobile phase and the cellulose-bound-water complex, dependent upon the solubility of the sugar in the eluant and the ease with which it can enter the structures of the complex and/or solid support. This latter ability is determined by its size and steric configuration. Generally cellulose t.l.c. has the same chromatographic characteristics as paper with the advantages that elution times are shorter and the sensitivity enhanced.

Silica gel separates in a similar manner but with an additional adsorption component. Often inorganic salts (e.g. phosphate) are impregnated into the gel by wetting with the salt solution followed by drying, after the plates have been coated, or by inclusion in the slurry solvent. In these cases, the selectivity of the inorganic salt greatly influences the carbohydrate separation and is, in turn, determined by its concentration and ionic form. In use, a gradient of salt is formed up the plate according to the composition of the eluant. An alternative to salt-impregnated silica gel is the use of non-impregnated silica gel in conjunction with a solvent containing phenylboronic acid. This system separates on the basis of the solubility and stability of the phenylboronic esters of the carbohydrates in the eluant, in addition to the straightforward partitioning effect of the underivatized carbohydrates. The presence of benzene and benzoates in the solvent have the greatest stabilizing effect on these esters.

Thin-layer plates can either be prepared in the laboratory or purchased ready-coated with cellulose, silica gel or phosphate-impregnated silica gel as the solid support. In general, pre-coated plates are to be preferred to home-made plates as they give excellent reproducibility, a higher sensitivity to detection reagents and, because of their bonded strength, they allow multiple elutions and reagent applications without their surface breaking up. Rapid and simple preliminary investigation of a system is possible by use of coated microscope slides. Clean dry slides are coated by dipping in a slurry of the chromatographic material, dried and run in a covered beaker.

The choice of a suitable solvent for t.l.c. is not easy unless a very simple sugar mixture is anticipated. The initial choice should lie between the six solvent systems suggested (*Tables 1* and *2*; solvents S1−S6) but if all of these prove unsatisfactory there is an abundance of choice in the literature (17−19). It should be noted, however, that even under optimal conditions a maximum of about 10 carbohydrates may be separated in a 1-dimensional run, increasing to about 20 if a suitable 2-dimensional system is appropriate. Carbohydrates with relative $R_f$s closer than about 5% cannot normally be resolved unless a discriminating detection system is used. Solvents are generally binary, tertiary or quaternary and always include an aqueous solution, preferably between 10 and 20% by volume. Small changes in the composition of such mixtures may have large and possibly unpredictable effects on the relative movement of the carbohydrate

**Table 1.** Solvent systems for use in t.l.c.

---

*S1: Ethyl acetate/pyridine/water*

Mix 100 ml of ethyl acetate with 35 ml of pyridine and 25 ml of water. This solvent is suitable for the analysis of hexoses, deoxyhexoses and some disaccharides on cellulose using three successive developments.

*S2: Butanol/pyridine/0.1 M HCl*

Mix 50 ml of *n*-butanol with 30 ml of pyridine and 20 ml 0.1 M hydrochloric acid (made up by adding 1 ml of concentrated HCl to 114 ml of water). This solvent is suitable for use on cellulose in order to separate monosaccharides derived from the acid hydrolysis of glycoproteins, i.e. galactose, mannose, fucose, glucosamine and galactosamine.

*S3: Formic acid/ethyl methyl ketone/tert-butanol-water*

Mix 30 ml of formic acid, 60 ml of ethyl methyl ketone (2-butanone), 80 ml of *tert*-butanol and 30 ml of water. This solvent is suitable for use on cellulose for the analysis of carbohydrates derived from plant extracts, including uronic acids. D-Arabinose may be distinguished from its L-isomer.

*S4: Ethyl methyl ketone/benzene/2-propanol/IPAB*

Dissolve 0.443 g of isopropylamine and 0.915 g of benzoic acid (IPAB) in 15 ml of water. Add 30 ml of ethyl methyl ketone (2-butanone), 40 ml of 2-propanol and 20 ml of benzene. See solvent system S5 for comments.

*S5: Ethyl methyl ketone/benzene/2-propanol/IPAB/PHEBA*

This is prepared as S4 (above) but with the addition and solution of 2.1 g phenylboronic acid (PHEBA).

These two solvent systems (S4 and S5) are recommended for use on silica gel thin layer plates. Certain carbohydrates (e.g. ribose, galactose and sorbose) exhibit increased $R_f$s in the presence of phenyl boronic acid whereas the $R_f$s of others (e.g. glucose, arabinose and fucose) are decreased. The two systems in combination are, therefore, useful for 2-dimensional t.l.c. System S4 is not recommended for use on its own (1-dimensional), however, as it does not give as good a resolution, in general, as the other solvent systems.

*S6: 2-propanol/acetone/0.1 M lactic acid*

Dissolve 148 mg of lactic acid in 20 ml of water. Add 40 ml of 2-propanol and 40 ml of acetone.

This solvent sytsem is recommended for use on phosphate impregnated silica gel plates. These are obtained by *either:*

(1)     preparing the coating silica gel slurry in 0.5 M $NaH_2PO_4$ or $KH_2PO_4$ (dissolve 19.5 g of $NaH_2PO_4.2H_2O$ in 250 ml of water), or
(2)     dampening pre-coated plates in 0.5 M $NaH_2PO_4$ followed by drying, or
(3)     buying phosphate-activated pre-coated silica gel plates.

   The system separates many of the more common constituents found in the clinical analysis of urine and plasma (e.g. glucose, galactose and mannose).

---

($R_f$) and the efficiency of the separations. Therefore, $R_f$ values should not be regarded as constant parameters, especially between laboratories. They may vary with temperature, humidity, coating batch, coating method and thickness, and chromatographic tank size. They should be established in the system under scrutiny and literature values should be used for guidance purposes only. As a general rule, carbohydrates have higher $R_f$s if they are more hydrophobic or of lower molecular weight. There are many apparent exceptions to this rule however.

## 3.2 Detection methods

There has been a large number of spray reagents described in the literature for the

**Table 2.** Retention data[a] and sample visualization[b].

| Monosaccharide | T.l.c. | | | | | | Paper chroma-tography | | Visualization | |
|---|---|---|---|---|---|---|---|---|---|---|
| | S1 | S2 | S3[c] | S4 | S5 | S6 | S7[d] | S8[c] | D1[e] | D2[e] |
| D-Xylose | 0.33 | –[f] | 1.34 | 0.74 | 0.74 | 0.63 | 1.00 | 2.5 | Faint grey blue | Brown |
| D-Ribose | 0.39 | – | 1.49 | 0.72 | 0.93 | 0.60 | 1.20 | 4.4 | Faint grey blue | Light blue |
| L-Arabinose | 0.31 | – | 1.30 | 0.57 | 0.51 | 0.47 | 0.76 | 2.3 | Faint grey blue | Light blue |
| D-Glucose | 0.25 | 0.53 | 1.00 | 0.51 | 0.47 | 0.41 | 0.56 | 1.0 | Blue grey | Violet |
| D-Galactose | 0.31 | 0.47 | 0.93 | 0.42 | 0.54 | 0.30 | 0.45 | 1.5 | Blue grey | Blue |
| D-Fructose | – | – | 1.21 | 0.53 | 0.63 | 0.46 | 0.77 | 1.7 | Light red | Purple/red |
| D-Mannose | 0.30 | 0.60 | 1.17 | – | – | 0.50 | 0.80 | 1.2 | Blue grey | Blue |
| L-Fucose | – | 0.71 | 1.44 | 0.70 | 0.66 | 0.65 | 1.10 | – | Olive intense blue | Pink |
| D-Glucosamine | – | 0.38 | – | – | – | 0.0 | – | – | Pale grey | Grey |
| D-Galactosamine | – | 0.29 | – | – | – | 0.0 | – | – | Pale grey | Grey |
| Reference | 17 | 25 | 19 | 18 | 18 | 24,49 | 23 | 22 | 17,49 | 17.49 |

[a]Values given are Rfs, except where indicated.
[b]Colours after incubation for the times indicated in the protocols listed in *Table 3*.
[c]$R_{glucose}$.
[d]$R_{xylose}$.
[e]Visualization reagent using silica gel t.l.c.
[f]Not reported.

detection of carbohydrates. Three are suggested here (*Tables 2* and *3*; reagents D1−D3) in order to cover most analyses. Two of these (D1 and D2) give different colours with different carbohydrates and are useful in identifying components in mixtures and overlapping unresolved carbohydrates. the third reagent (D3) is very sensitive with broad specificity. The colour produced by these detection systems is dependent on both the temperature and the duration of the heating period. This might affect the selectivity of the reaction. Additionally trace amounts of eluants persisting in the chromatographic layer may affect both the colour and sensitivity of the reactions. Care should be taken with pre-coated plates to avoid destroying the coating by over-heating during colour development.

It is possible to quantitate material separated by t.l.c. Direct densitometric assessment of the sprayed plates is possible but, due to difficulties arising from spot irregularity, perhaps should only be used with samples applied as bands in one-dimensional t.l.c. In addition, the spray detection techniques give poor reproducibility of quantitative results even on the same plate. It is best to locate the bands by use of co-chromatographed standards, identified by spray-detecting parts of the t.l.c. plates, while protecting other parts by shielding with a glass plate. The samples may then be scraped off from the unsprayed part of the plate, eluted and analysed by a specific colorimetric assay. Care must be taken that different impurities in standard or sample do not cause a difference in $R_f$.

### 3.3 General thin-layer chromatographic method

For 1-dimensional development, samples and standard mixtures (made by dissolving 0.05−2 mg dry component in 0.5 ml water plus 0.5 ml isopropanol) are applied, via a drawn-out capillary tube, as short streaks (1−2 cm long) 1.0 cm from the edge to

11

**Table 3.** Detection reagents for use in t.l.c.

*D1: Diphenylamine/aniline/phosphoric acid*

1. Prepare the following solutions:

   (A) Dissolve 4 g of diphenylamine in 80 ml of acetone and make up to 100 ml with more acetone.
   (B) Add 4 ml of aniline to 96 ml of acetone and mix well.
   (C) 85% Orthophosphoric acid.

2. For the spray reagent, mix 100 ml of A, 100 ml of B and 20 ml of C just prior to use.
3. Spray the plate, air dry and then heat at 100°C for 10 min. The colour appears after 2−4 min.

This spray reagent can be used for aldoses, ketoses, deoxysugars, oligosaccharides and uronic acids and may be used on both cellulose and silica thin-layer plates. It gives a wide variation of colours with different carbohydrates, aldoses producing blue grey spots whereas ketoses give light red spots. The sensitivity is about 1 $\mu$g.

*D2: Naphthoresorcinol/ethanol/sulphuric acid*

1. Prepare solution A by dissolving 0.2 g of naphthoresorcinol (naphthalene-1,3-diol) in 100 ml of 95% ethanol. Note that 0.4 g of diphenylamine may be added in order to reduce the background colouration of the chromatograms.
2. For the spray reagent, carefully add 4 ml of concentrated sulphuric acid to 96 ml of solution A just prior to use.
3. Spray the plate and then that at 100−150°C for 5 min.

This spray reagent can be used for aldoses, ketoses, uronic acids, deoxysugars, glycosides and oligosaccharides. It is recommended for use only on silica thin-layer plates. It gives a wide variation of distinctive colours with different carbohydrates; aldoses producing blue or violet spots, ketoses producing pink or red spots and uronic acids producing characteristic blue spots. The sensitivity is between 0.1 $\mu$g (L-sorbose) and 4 $\mu$g (D-glucose).

*D3: Tetrazole blue*

1. Prepare the following solutions:

   (A) Dissolve 0.2 g of tetrazole blue [3,3'-dianisole bis-4,4(3,4-diphenyl)tetrazolium chloride] in 100 ml of water.
   (B) Dissolve 4 g of sodium hydroxide in 80 ml of water and make up to 100 ml with more water.

2. For the spray reagent, mix 30 ml of solution A, 100 ml of Solution B and 50 ml of methanol just prior to use.
3. Spray the plate then heat at 40°C for 5−10 min.

This reagent can be used in all reducing and some non-reducing (e.g. sucrose) sugars. It may be used on cellulose and silica gel thin-layer plates and gives a simlar colour (Blue-violet) for all carbohydrates. The sensitivity is about 1 $\mu$g.

be dipped in the eluant and parallel to that edge. The volume applied should depend on the sensitivity of the detection system and the estimated concentration of the carbohydrate mixture. If more than one application is necessary, the previous application should first be dried well by a gentle draught of warm (not hot) air from a hand-held hair dryer. After sample application, the dry plates are placed in chromatographic tanks to which the elution solution has already been placed sufficient in advance to saturate its atmosphere. Filter paper placed along one side of the tank will help this process as with paper chromatography (see *Figure 1*). Given a choice, the coated face of the plate should be pointed towards the eluant rather than away from it. The top should be placed on the tank and the chromatogram developed at room temperature. This usually takes about 1 h/10 cm. If a multiple run is necessary (e.g. using a cellulose plate in

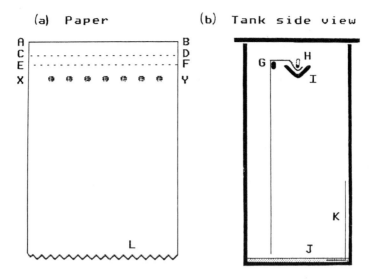

**Figure 1.** Paper chromatography. (**a**) Shows the chromatography paper ready for use. The samples should be applied as spots, no more than 5 mm in diameter, at least 2.5 cm apart along a line (X-Y) 8 cm from the top of the paper. The spots should be well dried in warm air before placing in the chromatography tank. Two folds are made in the paper, downwards along E-F and C-D (A-C ~3 cm, C-E ~2.5 cm, E-X ~2.5 cm). These folds enable the paper to be hung over the antisyphon rod (G) and secured under the support rod (H) in the top glass trough (I) as shown in the tank sideview (**b**). The atmosphere in the tank should be already saturated by the presence of some solvent in the bottom (J) aided by a small piece of chromatography paper (K) placed up one side. After the analytical paper has been well equilibrated in this atmosphere, solvent is added to the top trough (I) in order to start the chromatographic process. If the chromatogram is to be continuously developed, the bottom of the paper should have been cut with pinking shears (L), in order to ensure an even flow.

solvent S1 for hexoses), the plate should be removed and dried thoroughly in a stream of warm air before running again.

If a 2-dimensional development is to be attempted, the sample is applied as a spot to a corner position on the plate 1.0 cm from each side. The second run will be at right angles to the first in a significantly different solvent. The plate should be thoroughly dried in a stream of warm air before the second run is commenced. A difficulty with this method concerns the standards since they have to be included with the unknown sample and co-chromatographed. Urea is reported to be a useful internal standard as it is generally well separated from carbohydrates but can be usually identified with the naphthoresorcinol reagent (*Table 3*; reagent D2). The plates should be thoroughly dried in a stream of warm air before using the spray reagents for detecting the sugars. It is suggested that a commercial aerosol spray gun be used, spraying into a fume cupboard.

High performance t.l.c. pre-coated plates are now available from a number of suppliers. These have a much narrower particle size and particle size range enabling an improved resolution at higher efficiency. In addition more samples may be applied due to the small degree of spreading and, hence, spot size. The smoother, more homogeneous, media gives improved reproducibility and samples may be quantitated by densitometry in a far more satisfactory manner than standard t.l.c. plates. Some of these plates also come with a concentration zone for improved sample application. The major and only drawback to the use of these plates is their relatively high cost.

**Table 4.** Solvent systems for use in paper chromatography.

---

*S7: n-Butanol/pyridine/water*

Mix 100 ml of *n*-butanol with 30 ml of pyridine and 30 ml of water. (Pazur, see Chapter 3, recommends a solvent consisting of 60 ml of *n*-butanol, 40 ml of pyridine and 30 ml of water for use with polysaccharide hydrolysates.) This solvent is suitable for the analysis of carbohydrates derived from glycoproteins (21). A run takes about 25 h at room temperature. The chromatogram should be dried in a fume cupboard for about 4 h (until only a faint odour of *n*-butanol remains) before use of dip detection reagents.

*S8: Nitromethane/acetic acid/ethanol/boric acid[a]*

Mix 160 ml of nitromethane with 20 ml of glacial acetic acid, 20 ml of ethanol and 20 ml of a saturated aqueous boric acid solution. The chromatogram should be run at a constant temprature of about 40°C.

---

[a]From ref. 22.

## 4. PAPER CHROMATOGRAPHY (19,20)

### 4.1 Experimental approach

Paper chromatography has similar chracteristics to t.l.c. on cellulose. It is mainly a partition process with the possibility of some adsorption. It is a cheap and simple method and easier to use, on a preparative scale than t.l.c. The use of dip reagents for visualizing the spots (see Section 4.2) is also an easier and more uniform process than spraying. Although the solvents, developed for cellulose-coated t.l.c. may be used in paper chromatography, some systems have been developed particularly for the paper medium. Two such solvents are described in *Tables 2* and *4* (solvents S7 and S8). In general, the preferred conditions are descending chromatography on Whatman No.1 paper (*Figure 1*). Multiple or continuous development may be used. For this latter technique pinking shears are used on the bottom end of the paper to ensure a uniform flow of the solvent. Thicker paper (e.g. Whatman No.17) can be used for preparative separations of up to about 0.5 g, bands being visualized by using a dip reagent on a copy of the chromatogram 'blotted' onto thin chromatographic paper (e.g. Whatman No.1). Chromatography paper is most often used straight from the pack but prewashing with 0.25% (w/v) hydroxyquinoline in 8% (w/v) acetic acid may be beneficial. In either case the paper should be well equilibrated in the solvent vapours for several hours before use. For reproducible $R_f$s the chromatographic runs should be performed under constant temperature conditions in cold (slower running) or warm (faster running) rooms. Many a run has been spoilt by running overnignt in a laboratory with no overnight heating.

### 4.2 Dip reagents for paper chromatography

Spots on paper chromatograms may be visualized by means of 'dip' reagents. These are placed in a wide shallow dish and the dried chromatogram, held at each end, is drawn rapidly but smoothly through the reagent. This method is easier and more reproducible than spraying, as used in t.l.c. The wet chromatograms are carefully clipped to a line or glass bar in a fume cupboard in order to dry. A certain amount of forethought must be put into this procedure as the chromatogram may be fairly fragile in this wet state. Although the spray reagents developed for cellulose-coated t.l.c. may also be used for paper chromatography, some detection systems have been developed particularly for the paper medium. Three such dip detection reagents are described in *Table 5* (reagents D4−D6).

**Table 5.** Dip reagents for use in paper chromatography.

*D4: Alkaline silver nitrate*

1.  Prepare the following solutions:

    (A)   Add 0.1 ml of a saturated aqueous solution of silver nitrate to 20 ml of acetone. If a cloudiness appears, add a drop more water.

    (B1)  Dissolve 2 g of sodium hydroxide in a minimal volume of water (<1 ml) and make up to 100 ml with methanol.

    (B2)  Dissolve 2 g of sodium hydroxide and 4 g of pentaerythritol in 20 ml of water. Add 80 ml of ethanol and mix throughly.

2.  Dip the dried chromatogram in reagent A, dry again and then spray with reagent B1. If the boric acid containing solvent system S8 is used for the chromatographic development then reagent B1 should be replaced by reagent B2.

This is a sensitive general stain producing silver-brown spots with reducing carbohydrates at a sensitivity of about 1 $\mu$g. Non-reducing compounds do not produce the colour at once and may require a short heating period.

*D5: Elson−Morgan reagent*

1.  Prepare the following solutions:

    (A)   Dissolve 5 g of KOH in 20 ml water and add 80 ml of 95% ethanol.

    (B)   Mix 1 ml of redistilled pentane-2,4-dione (acetylacetone) with 99 ml of 95% ethanol immediately before use.

    (C)   Dissolve 10 g of *p*-dimethylaminobenzaldehyde in concentrated hydrochloric acid.

    (D)   95% ethanol.

2.  Dip the dried chromatograms in a mixture of reagents A and B (1 volume A plus 10 volumes B), dry and heat at 110°C for 5 min. Dip the partially reacted chromatograms in a (1:1 v/v) mixture of reagents C and D. On drying in cold air, transient purple spots indicate aminodeoxyhexoses. Heating to 80°C gives permanent red spots for these aminodeoxyhexoses and purple-violet spots for the N-acetylated derivatives. Excessive heating should be avoided or else the brown background will mask these spots. The sensitivity for this reagent is about 1 $\mu$g. It should be used where confirmation of the presence of aminodeoxyhexoses and their N-acetylated derivatives is required.

*D6: Diphenylamine/aniline/phosphoric acid*

1.  Dissolve 0.15 g of diphenylamine in 25 ml of ethyl acetate. Add 0.8 ml of aniline and 75 ml of ethyl acetate. Finally add 1 ml of water and 10 ml of concentrated phosphoric acid. The reagent should be made up immediately before use.

2.  The dry chromatograms should be dipped through the reagent, dried and heated at 95−100°C until the background is faintly grey. This is a good general purpose reagent giving similar results to the spray reagent D1.

## 5. HIGH PERFORMANCE LIQUID CHROMATOGRAPHY

The separation and quantitation of the components in mixtures of monosaccharides forms an important part of carbohydrate analysis. There are two main methods available, h.p.l.c. and g.l.c. The choice between these two depends on a number of factors in addition to personal preferences and prejudice. One clearly important factor is the available resources and expertise. Both methods need fairly expensive equipment and are facilitated by helpful and experienced operators. Experience in running these systems can only be built up over a fairly long period through a number of mishaps and, therefore, it is best if it is gained second-hand. As a broad generalization, h.p.l.c. is preferred for the analysis of simple monosaccharide mixtures and oligosaccharide analysis and

**Table 6.** Typical conditions for the h.p.l.c. of monosaccharides.

| Column type | Mobile phase | Temperature (°C) | Flow rate (ml/min) | Analysis | Commercially available columns (i.d. × length) (cm) |
|---|---|---|---|---|---|
| Anion exchange (quaternary ammonium, borate form) | Borate buffers | 65 | 1.0 | Alditols, mono-saccharides, sialic acids | Hitachi No. 2633 (0.8 × 8)<br>Durrum DA-X8-11 (0.3 × 25)<br>Bio-Rad Aminex A-14 (0.9 × 99)<br>Bio-Rad Aminex A-25 (0.4 × 25) |
| Anion exchange (amino, OH⁻ form) | Acetonitrile/water | 25 | 2.5 | Monosaccharides, oligosaccharides | Waters μBondapak Carbohydrate (0.4 × 30)<br>Whatman Particil 10-PAC (0.4 × 25)<br>Varian MicroPak NH$_2$ (0.4 × 30)<br>Merck Lichrosorb NH$_2$ (0.4 × 25)<br>Supelco Supercosil LC-NH$_2$ (0.46 × 25) |
| Cation exchange (sulphonate, H⁺ form) | Acetonitrile/water | 30 | 0.6 | Glycoprotein-derived carbohydrates | Shodex DC-613 (0.4 × 25)<br>Bio-Rad Aminex HPX-87H (0.78 × 30) |
| Cation exchange (sulphonate, Ca²⁺ form) | Water | 85 | 0.6 | Alditols, mono-saccharides | Bio-Rad Aminex Q-15S (0.9 × 50)<br>Bio-Rad Aminex HPX-87C (0.78 × 30)<br>Waters Sugar-Pak 1 (0.4 × 30) |
| Silica, straight phase | Acetonitrile/water | 25 | 0.54 | Pre-derivatized carbohydrates | Whatman Partisil 0 (0.46 × 25)<br>Merck (LiChrosorb Si60 (0.4 × 25)<br>Waters μ Porosil (0.4 × 30) |
| Silica, straight phase | Acetonitrile/water + diaminoalkanes | 25 | 1.0 | Monosaccharides, oligosaccharides | Whatman Partisil 0 (0.46 × 25)<br>Merck (LiChrosorb Si60 (0.4 × 25)<br>Waters μ Porosil (0.4 × 30) |
| Silica, reverse phase | Acetonitrile/water | 28 | 1.0 | Pre-derivatized carbohydrates | Waters μBonakpak (C$_{18}$) (0.39 × 30)<br>Merck LiChrosorb RP-18 (0.4 × 25)<br>Waters Radialpak C$_{18}$ (0.4 × 30)<br>Supelco Supelcosil LC-18 (0.45 × 25)<br>DuPont Zorbax ODS (0.46 × 25) |

purification whereas g.l.c. can be used on very complex monosaccharide mixtures. There are, however, protocols available for separating and analysing most mixtures of carbohydrates by either method. Advances in both methods are constantly being made and reported in the literature (e.g. in Analytical Biochemistry and Journal of Chromatography) and by the column manufacturers and suppliers. This section describes the practical use of h.p.l.c. for carbohydrate analysis and Section 6 describes g.l.c.

## 5.1 **Experimental approach**

There are a number of different h.p.l.c. processes for separating carbohydrates which depend on different chemical and physical properties for resolution (see ref. 45 and *Table 6*). This is a somewhat confusing state of affairs but no single method is useful over the entire range of possible separations. In general terms, these analyses can be divided into ion exchange processes, at high temperature ($\geq 60°C$) and adsorption (partition) processes at lower (room) temperatures. The most commonly used ion-exchange process involves the separation of borate complexes on quaternary ammonium anion exchange resins (see ref. 28 and *Figure 2*). Borate forms complexes with most carbohydrates. The stability of these is influenced by the spatial arrangement of the alcohol groups involved, *cis* hydroxyl groups forming the most stable compounds. The columns may be run isocratically using borate buffer or by use of gradient or stepwise elution using borate buffers of increasing molarity (although not necessarily increasing

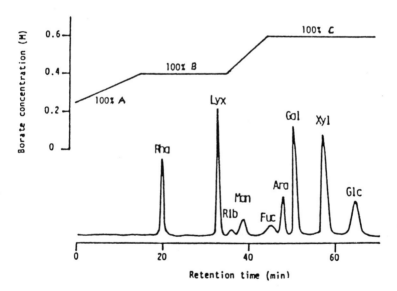

**Figure 2.** Analysis of an equimolar mixture of aldoses (28). A mixture of L-rhamnose (30 nmol), D-lyxose (30 nmol), D-ribose (60 nmol), D-mannose (30 nmol), L-fucose (60 nmol), L-arabinose (30 nmol), D-galactose (30 nmol), D-xylose (15 nmol) and D-glucose (30 nmol) were dissolved in 100 μl of water and applied to a Hitachi No. 2633 resin column (8 mm i.d. × 8 cm). The packing is a quaternary ammonium resin with an average bead diameter of 11 μm. The borate gradient (1 ml/min) was used to elute the carbohydrates using buffers A (0.25 M, pH 8.2), B (0.4 M, pH 7.4) and C (0.6 M, pH 9.3). Post column detection was by fluorimetry after reaction width 2-cyanoacetamide/borate [see post-column detection method, Section 5.2.3 (P3)]. Redrawn from ref. 28 with permission.

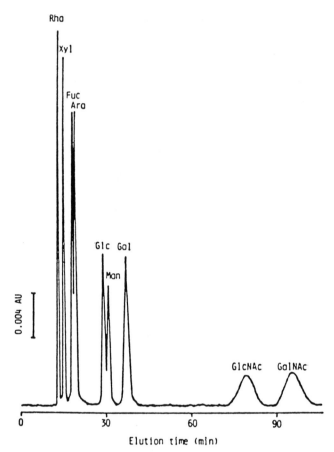

**Figure 3.** Analysis of an equimolar mixture of aldoses and *N*-acetyl hexoamines. Five nmol each of a mixture of L-rhamnose, D-xylose, L-fucose, L-arabinose, D-glucose, D-mannose, D-galactose, *N*-acetyl-D-glucosamine and *N*-acetyl-D-galactosamine were dissolved in 20 $\mu$l water:acetonitrile (8:92) and applied to a Shodex DC-613 column (4 mm i.d. × 25 cm). The column was a highly crosslinked sulphonated polystyrene resin in the H$^+$ form which separates by partition chromatography. It was run isocratically at 30°C, 0.6 ml/min, using aqueous 92% v/v acetonitrile with post column detection by u.v. 280 nm absorbance after reaction with 2-cyanoacetamide/borate [see post-column detection methods, Section 5.2.3 (P3)]. This system can be used for the analysis of carbohydrates derived from glycoproteins. The amino-sugars produced by hydrolysis must be reacetylated (see Section 6.2.1) before analysis. *N*-Acetyl and *N*-glycollyl neuraminic acids may be analysed (as *N*-acylmannosamines) after hydrolysis and conversion using *N*-acetylneuraminidase (0.5 U) and *N*-acetylneuraminate pyruvate-lysate (0.3 U, 1 h 37°C, 200 $\mu$g glycoprotein in 800 $\mu$l 0.06 M phosphate buffer pH 7.0). From ref. 29 with permission.

pH) or containing increasing salt concentrations. It is not normally necessary to regenerate the column after each run as re-equilibration with the starting buffer is sufficient preparation. The peaks produced tend to broaden with the elution time but excessive backpressure prevents a significant increase in the flow rate. The resolution of this process normally increases with increasing temperature as long as the ion-exchange resin is stable. Cation exchange resins may be used directly for the separation of amino sugars or, by use of the calcium form (ligand-exchange chromatography), most alditols and monosaccharides. Separation processes involving partition may make use of a number of different types of stationary phase: amino-derivatized anion exchange resins, cation exchange resins ($H^+$ form) (see ref. 29 and *Figure 3*) and physicially derivatized silica. Monosaccharides may be separated in all these processes using aqueous acetonitrile solvents. Derivatized carbohydrates may be analysed by standard techniques using straight phase or reverse phase silica columns. Some stationary phases are available in 'radial compressed' columns where the inside surface of the walls conform to the outer contours of the particles within the bed when under compression. This method compacts and stabilizes the column, eliminates voids, reduced 'edge effects' and improves column reliability.

Optimization of the conditions for the h.p.l.c. separation of a number of known carbohydrates is a straightforward if somewhat lengthy and tedious process. The variation of elution times of all the components with the separation parameters (e.g. temperature, acetonitrile concentration, buffer pH) must be determined. The optimum conditions may then be determined by computer analysis of the resultant curves or by visual inspection of the curves generated for the resolution of all pairs of components against the separation parameter chosen (see refs. 28 and 29 for further details).

The columns used in h.p.l.c. are usually bought ready-packed. They may be packed in-house, however, with a saving in cost and a possible increase in efficiency. The best method appears to be slurry-packing. The calculated amount of the packing material is dispersed in the slurry medium (often water) by use of an ultrasonic bath and thoroughly degassed. The slurry is then pressed into the column by use of the expected elution solution at a high enough pressure to achieve at least twice the analytical flow rate. In use, the columns gradually deteriorate. This process of deterioration can be reduced by careful attention to procedure. All solvents to be passed through the column should be as pure as possible and have been filtered through a 0.5 $\mu$m filter, to avoid plugging of the frits by fine particles, and thoroughly degassed, preferably by use of helium bubbled through air stones. Where possible, microbial growth inhibitor (e.g. 0.01% w/v sodium azide) should be present in all aqueous phases. Samples applied to columns should also be free from particulate matter and other gross contaminants (e.g. proteins). The analytical column should, where feasible, be protected from contamination by use of a short 'guard' column containing the same type of resin (although possibly with a larger bead size). It should also be used only within the manufacturer's recommended range of pH and solvent. This is particularly important for silica based stationary phases which dissolve readily at alkaline pH. At the end of each analytical session the column and all eluant delivery lines should be rinsed through with pure solvent (e.g. water) especially if salt solutions were being used. Where the column is constantly being removed from the h.p.l.c. apparatus it should be stored carefully and not dropped. Even with all these precautions some alkylated silica columns have a fairly short life of as little as 3 months due to particle dissolution and loss of the bonded phase. If, however, the analytical method fails suddenly, suspect the sample preparation before the column.

## 5.2 Detection methods

### 5.2.1 *Direct detection*

The direct detection of carbohydrates is not straightforward as they do not absorb light in the normal u.v. or visible range and have no fluorescence. Monosaccharides, however,

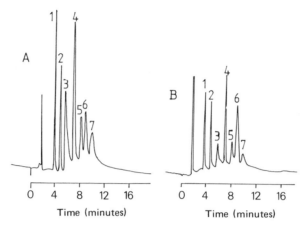

**Figure 4.** Separation of monosaccharides by h.p.l.c. 200 μg of L-rhamnose (1), D-xylose (2), D-arabinose (3), D-fructose (4), D-mannose (5), D-glucose (6) and D-galactose (7) were separated on a Supelcosil LC-NH₂ column (25 cm × 4.6 mm i.d.) using acetonitrile:water (17:3 v/v) as mobile phase at 2 ml/min. The eluates were detected by absorption at 190 nm (**A**) or by use of refraction index detector (**B**).

do absorb at wavelengths in the far u.v. The absorbance maxima are at about 188 nm but, due to noise in the detection signal below 190 nm, detection is normally performed at wavelengths between 192 nm and 200 nm. The response, depending largely on the freedom of the carbonyl group, differs between the monosaccharides, being six times higher for D-fructose (~2 μg in 20 μl, 0.6 mM) than D-glucose; the response of most other carbohydrates lying between these extremes (*Figure 4A*). Analyses are restricted to solvents which do not absorb significantly. The response is linear over a wide range but the absolute response depends on other parameters such as the solvent purity, the reference cell and the spectrophotometer's condition. Detection in this range requires an expensive instrument and a very pure solvent. Acetonitrile/water mixtures are often used. A reliable supplier of non-absorbing h.p.l.c. grade acetonitrile should be sought as this has often been found to be the critical component and it is not easy or practicable to purify in-house.

Another detection method of general applicability makes use of changes in refractivity. The refractive index detector is inherently as sensitive as absorbance measurement in the far u.v. but, in practice, is at least ten times less sensitive due to the background noise caused by pump pulsations and temperature fluctuations. The method is generally applicable to carbohydrate detection but is not suitable for separations involving gradient elution. The sensitivity of detection varies between carbohydrates and with the solvent composition but is generally linear (*Figure 4B*).

### 5.2.2 *Pre-column derivatization*

In order to increase the sensitivity of detection of carbohydrates, they may be derivatized to give light-absorbing or fluorescent materials before or after separation on the h.p.l.c. column. Pre-column derivatization not only increases the carbohydrate sensitivity to detection but also substantially changes its chromatographic behaviour. Suitable procedures are given in *Table 7*. This approach may be very useful, especially if it allows the use of the highly efficient derivatized (e.g. C₁₈ reverse phase) or underivatized silica columns (*Tables 6* and *7*).

**Table 7.** Pre-column derivatization of carbohydrates for use in h.p.l.c.

| Reagent | Sensitivity[a] | Column | Eluant | Reference |
|---|---|---|---|---|
| Benzoyl chloride | 2 | Silica | Ethylacetate:hexane | 30 |
| Aminopyridine | 2[b] | C18 reverse phase | Acetonitrile:citrate buffer | 31 |
| Dansyl hydrazine | 2[b] | C18 reverse phase | Acetonitrile:water | 32 |
| Phenylisocyanate | 3 | C18 reverse phase | Acetonitrile:phosphate buffer | 33 |
| 4-Bromobenzoyl chloride | 1 | Silica | Dichloromethane: ethyl acetate:hexane:diethyl ether | 34 |

[a]ng of glucose, as determined by u.v. absorbance.
[b]by fluorescence.

**Table 8.** Post-column detection of specific classes of carbohydrates.

| Carbohydrate | Assay | Sensitivity (nmol) | Reference |
|---|---|---|---|
| Diols | Periodate oxidation | 1 | 35 |
| Neutral carbohydrates | Orcinol-sulphuric | 5 | 41 |
| Reducing sugars | 4,4'-Dicarboxy-2,2'-biqinoline | 10 | 42 |
| Amino sugars | *o*-Phthalaldehyde | 0.04[a] | 43 |
| Sialic acids | Thiobarbituric acid | 1.5 | 44 |

[a]Fluorimetric assay: all the others involve colorimetric assay.

### 5.2.3 Post-column derivatization

In general, post-column derivatization should be the method of choice if a greater sensitivity than that achieved by direct detecton is required, since using this approach the separation process can make full use of the differences between the carbohydrates without the addition of a number of groups with intrinsically similar properties. Post-column derivatization is, in general, a straightforward technique requiring one or two additional pumps, a mixing coil of Teflon tubing and a thermostatted bath. Three generally applicable but simple and sensitive methods are fully described below (P1−P3) and a number of other more specific methods are given in *Table 8*. Quantitation from the detector's response is made by direct comparison with standards (or standard curves). The amounts are usually proportional to the area of the peak in the detector output. This may be measured by use of an integrator or, more cheaply, by physically weighing on a balance the peak cut out from a photocopy of the trace. This method is particularly useful for overlapping peaks. If the system gives reproducible elution times, the peak heights are normally proportional to the amount of material for any given compound.

### 5.2.4 P1: Ammoniacal cupric sulphate (26)

*Sensitivity:* 450 ng glucose in 20 $\mu$l (125 $\mu$M).

*Reagent:* Prepare a stock solution by dissolving 25 g of cupric sulphate in 500 ml of water and then adding 500 ml of concentrated ammonia solution (35% w/v $NH_3$, sp. gr. 0.88). Prepare the working solution from the stock by diluting 100 ml with 250 ml of the concentrated ammonia solution and 650 ml of water. Filter the solution through a 0.2 $\mu$m pore size glass fibre filter before use.

*Method.* Mix the column eluate with the reagent by means of a Y shaped joint at the column outlet. The flow rates of the eluate and reagent should be approximately equal, although a slight excess of reagent may be beneficial. The reaction is immediate. The absorbance may be measured between 285 nm, where the assay gives a more sensitive response, and 310 nm, where there is a lower baseline absorbance. The choice of wavelength may depend on the characteristics of the spectrophotometer used as a very stable performance is necessary at the lower wavelengths, due to the high background absorbance. Essentially pulse-free delivery of the reagent is important in order to attain a steady baseline. Waste from the assay should be pumped into a bottle vented to a sink or fume cupboard. The flow cell should be washed through regularly (e.g. weekly) with 20% nitric acid.

*Comments.* The assay is sensitive but simple and avoids the corrosive reagents and comlex heating/mixing protocols of some other post-column detection methods. It is of general application to substances that react with cuprammonium, for example, carbohydrate derivatives and glycols, and is not sensitive to changes in the solvent composition.

### 5.2.5 *P2: Alditol assay* (27)

Sensitivity: Photometric; 400 ng of glucitol in 10 $\mu$l (200 $\mu$M).
               Fluorimetric; 90 ng of glucitol in 10 $\mu$l (50 $\mu$M).

*Reagents:*

(A)    Dissolve 10.7 g of sodium metaperiodate in 1 l of water.
(B)    Dissolve 150 g of ammonium acetate and 57.6 g of sodium thiosulphate pentahydrate in 980 ml of water. Add 20 ml of acetylacetone and mix well.

*Method*

(i)    Mix the column eluate (1 ml/min) with reagent A (0.5 ml/min) via a Y shaped joint. Allow the periodate oxidation to proceed at room temperature for 80 sec by use of a 10 m open tubular Teflon coil (0.5 mm i.d.).
(ii)    Add reagent B (0.5 ml/min) via another Y shaped joint. Pass the mixture through another 10 m open tubular Teflon coil (0.5 mm i.d.) held in a thermostatted bath (100°C) containing glycerol. This reaction takes about 1 min.
(iii)    Determine the product spectrophotometrically at 412 nm or fluorimetrically at 410 nm (excitation)/503 nm (emission).

*Comments.* Monosaccharides do not seriously interfere in this assay if the periodate oxidation reaction takes place below 40°C and they are not present in an overwhelming excess. They can be detected, however, if the periodate oxidation is allowed to take place at a higher temperature (e.g. ~100°C). Periodate oxidation alone may be used for post-column detection of cyclitols, aldoses, alditols and ketoses by monitoring the absorbance at 260 nm. The reagent should be 0.1 mM $KIO_4$ in 0.5 M potassium borate buffer, pH 8.5 [see P3 (reagent B)] using a 5 min heating period at 90°C (35).

### 5.2.6 *P3: 2-Cyanoacetamide* (28,29)

*Sensitivity:* Photometric; 18 ng glucose in 20 $\mu$l (5 $\mu$M).

Fluorimetric; 2 ng glucose in 20 $\mu$l (0.5 $\mu$M).

*Reagents:*

(A) Dissolve 2-cyanoacetamide in methanol, decolourize with activated carbon and recrystallize from methanol. Dissolve 10 g of the purified 2-cyanoacetamide in 100 ml of water. This solution may be stored in a refrigerator within a dark bottle for at least one month.

(B) Dissolve 31 g of boric acid in 600 ml of water. Make the solution pH 8.5 by the addition of 10% (w/v) potassium hydroxide and dilute with water to a final volume of 1 litre (0.5 M borate).

*Method*

(i) Mix the column eluate (0.6 ml/min; if a different elution rate is used, scale all the rates accordingly) successively with reagent A (0.5 ml/min) and reagent B (0.5 ml/min) by use of Teflon Y shaped joints.

(ii) Pass the mixture through an open tubular Teflon reaction coil (10 m × 0.5 mm i.d.) held at 100°C in a thermostatted bath containing glycerol (80 sec contact time) followed by an air-cooled Teflon cooling coil (1 m × 0.5 mm i.d.).

(iii) Determine the absorbance at 280 nm (275 nm maximum) or the fluorescence at 331 nm (excitation)/383 nm (emission).

*Comments.* The assay uses non-corrosive reagents, shows good linearity and is highly sensitive for aldoses, hexosamines and alduronic acids. The fluorescence is quenched by acetonitrile, if present in the eluate, but the absorbance is unaffected.


## 6. GAS-LIQUID CHROMATOGRAPHY

### 6.1 **Experimental approach**

G.l.c. is preferred to h.p.l.c. in a number of instances. It is a more sensitive technique, allowing the analysis of sub-nanomolar amounts of carbohydrates, and is generally less prone to interference (e.g. from salts and protein). Detection is usually by means of a flame ionization detector (FID) which responds to all carbohydrate related molecules over an extremely wide linear range. G.l.c. separation is dependent upon the differential extractive distillation of the components in the mixture. It is fundamental to the technique, therefore, that volatile derivatives of the carbohydrates are prepared. There are two schools of thought concerning this derivatization. One is to produce a single derivative from each carbohydrate and the other allows several derivatives to be produced dependent upon the anomeric composition. The advantage of the former is that the chromatograms are simpler and the sensitivity may be marginally greater. However by producing more than one derivative in a well defined mass ratio, a recognizable 'fingerprint' can be obtained which is confirmatory for a particular component. This reduces the possibility of a chromatographic peak, due to an unexpected impurity, be-

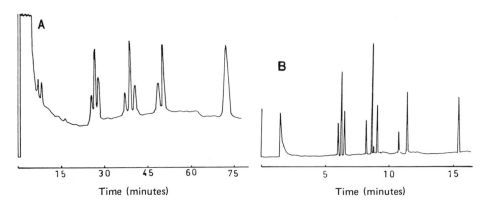

**Figure. 5.** Comparison of the separation efficiency of (**A**) a packed bed column (1.5 m × 6 mm i.d. 3% SE-30 on 100/200 mesh Gas Chrom Q) and (**B**) a WCOT column (25 cm × 0.32 mm i.d. coated with CP sil 5 liquid phase). The mixture consists of the per-trimethylsilyl derivatives of the methyl glycosides of L-fucose, D-galactose, *N*-acetyl-D-galactosamine and *N*-acetyl-neuraminic acid.

ing incorrectly assigned to a carbohydrate. With the commonplace use of the highly efficient, fused-silica, wall coated open tubular (WCOT) columns, it is no longer as important to simplify analyses of complex mixtures to one peak/component as the separation system is generally powerful enough ($\geq 100\ 000$ theoretical plates) to resolve a large number of components ($>30$). Although much of the data concerning g.l.c. separations has been derived from work using packed columns (19,46), the much better resolution ($>10$-fold) and speed ($>5$-fold) of WCOT columns recommends them for routine analytical work (*Figure 5*). There is no substantial change in the relative elution positions between these two types of column if similar liquid phases are used (*Table 9*). WCOT columns are more fragile and deteriorate more easily than packed bed columns. In general, this deterioration may be greatly reduced by the use of oxygen-free dry helium as the carrier gas. This helium should be passed through the appropriate gas purifiers before entering the column as oxygen has a particularly devastating effect on the column. If well-treated, and run below the supplier's maximum operating temperature limit, the WCOT columns should last for over a year of continuous use, which is longer than most h.p.l.c. columns last.

Internal standards are generally necessary for g.l.c. analyses since it is difficult to inject a reproducible proportion of a sample into the chromatograph. Standards may be any similar compound which is not already present in the mixture to be analysed and which is clearly separable from the other components. *Myo*-inositol (mesoinositol) may be used, except for plant or membrane carbohydrate analyses where it is usually already present. An alternative internal standard is pentaerythritol. Glucose should be avoided as an internal standard, even if it is known not to be expected in the samples, as it is a common, and annoying, contaminant.

## 6.2 Detection methods

Two g.l.c. detection methods are described below (D1 and D2). The former, involving trimethylsilylation, is a good, generally applicable, method whereas the latter, involving acetylation, is recommended particularly for methoxy-derivatized

**Table 9.** Alternative liquid phases for the g.l.c. of carbohydrates.

| Polarity | Typical composition | Alternative phases |
|---|---|---|
| Most nonpolar | Dimethyl silicone | SE-30<br>OV-1<br>OV-101<br>CP SIL 5CB |
| Nonpolar | 5% Phenyl, methyl silicone | SE-52<br>SE-54<br>CP SIL 8CB |
| Intermediate polarity | 50% Phenyl, methyl silicone | OV-17 |
| Polar | 50% Trifluoropropyl, methyl silicone | OV-202<br>OV-210<br>QF-1 |
| Most polar | Diethylglycol adipate or cyanopropylmethyl, phenylmethyl silicone | OV-275<br>CP SIL 88<br>ECNSS-M<br>Silar 9CP[a]<br>Silar 10CP[a]<br>LAC-IR-296 |

[a]Silar 10CP is more polar than Silar 9CP.

monosaccharides derived from the methylation of more complex molecules (e.g. glycoproteins, polysaccharides or glycolipids). Quantitation of the components in a mixture is by analysis of the peak areas (heights) as described for h.p.l.c. (Section 5.1).

### 6.2.1 *D1: Preparation of trimethylsilyl derivatives* (36)

*Reagents:*

(A) Dry methanol. It may be possible to purchase this. Otherwise methanol may be dried by refluxing 500 ml with 2.5 g of magnesium turnings and 0.1 g of iodine for 1 h, followed by distillation into a clean dry container.

(B) Anhydrous pyridine, bought or prepared by the addition of 30 g KOH pellets to 500 ml of pyridine. The mixture should be well stirred and allowed to settle for 24 h before use.

(C) Methanolic hydrogen chloride. This may be prepared by either of two methods.

    (i) Bubble hydrogen chloride gas through a train of moisture traps, containing concentrated sulphuric acid, dry methanol (reagent A) until the weight of the methanol solution increases by about 3%. Care should be taken over preventing the sulphuric acid traps sucking back when the process is finished. The methanolic HCl should be standardised to 0.625 M by the titration of an aliquot against standard 0.1 M sodium hydroxide using phenolphthalein as an indicator, and dilution with more dry methanol.

    (ii) Add 4.65 ml of good quality acetyl chloride carefully to 100 ml of dry methanol.

(D) Trimethylsilylation reagent. This may be purchased, ready-mixed and ampouled, but can be prepared by mixing dry pyridine (10 vol) with hexamethyldisilazane

**Table 10.** Suitable carbohydrate standard mixtures for g.l.c. analysis.

| Monosaccharide | Analysis[a] | |
| --- | --- | --- |
| | Glycoprotein (mg) | Plant sugars (mg) |
| L-Fucose | 5 | – |
| D-Mannose | 2 | 2 |
| D-Galactose | 2 | 2 |
| D-Glucose | 2 | 2 |
| N-Acetyl-D-glucosamine | 5 | 5 |
| N-Acetyl-D-galactosamine | 5 | 5 |
| N-Acetyl neuraminic acid[b] | 5 | – |
| L-Arabinose | – | 5 |
| L-Rhamnose[c] | – | 5 |
| D-Xylose | – | 3 |
| D-Glucuronic acid | – | 5 |
| D-Galacturonic acid | – | 6 |

[a]The mixtures should be made up from previously dried monosaccharides. Stock solutions made from these mixtures may be diluted and aliquotted according to the analytical method chosen.
[b]Contains acetic acid of crystallization which prevents the use of $P_2O_5$ for drying since it causes decomposition. For accurate work this should be standardized but no suitable independent method exists.
[c]Usually supplied as the monohydrate.

(2 vol) and trimethylchlorosilane (1 vol). It is best stored in sealed or serum capped ampoules under nitrogen. An alternative reagent for use on carbohydrate syrups can be prepared by mixing dry pyridine (10 vol) with hexamethyldisilazane (9 vol) and good quality trifluoroacetic acid (1 vol). This may be used in the presence of up to 2.0 mg water/ml reagent if the carbohydrate content is fairly high ( ~ 1 mg/ml reagent).

*Method.* A dry cabinet (glove box) should be used throughout for all reagent manipulations.

(i)    Dry the samples containing glycoproteins ( ~ 0.5 – 100 μg), polysaccharide or monosaccharide mixtures or standards ( ~ 0.1 – 100 nmol, *Table 10*) together with *myo*-inositol ( ~ 100 ng; or other suitable internal standard) over $P_2O_5$, under vacuum, in 100 μl or 300 μl Reacti-Vials (screw topped, serum capped vials, e.g. as supplied by Pierce).

(ii)   Add a mixture of dry methanolic HCl (4 vol) and methyl acetate (1 vol, total volume = ~ 40 – 200 μl depending on the expected carbohdyrate content). If reagent C(ii) is used (see above), which already contains some methyl acetate, the ratio of methanolic HCl to additional methyl acetate should be 5:1 (v/v). Seal the tube by means of the Teflon-lined septa in the screw caps and mix throughly be use of a vortex mixer.

(iii)  Incubate the tubes at 70°C in an oven for about 16 h. Check the screw caps for tightness after the first 10 min of this incubation and vortex mix the tubes again before replacing in the oven. Only very occasionally will this system of sealing fail ( ~ 1 in 50), usually due to use of a cracked or chipped tube.

(iv)   After cooling, add redistilled 2-methyl-2-propanol (*t*-butyl alcohol, 20% v/v).

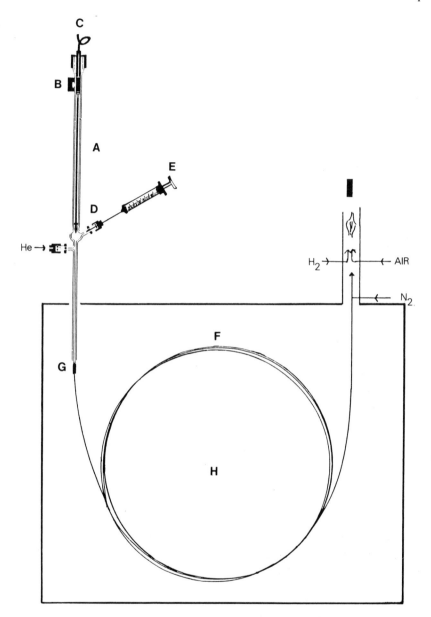

**Figure 6.** Analytical arrangement for the g.l.c. of per-trimethylsilylated methyl glycosides. The injector consists of all glass needle (A) held within a glass tube by means of a magnet (B) allowing the flow of helium gas over its surface to a restricted vent (C). The glass tube has a serum capped port (D) for the injection of samples via a syringe (E) and connected to the 25 m × 0.32 mm i.d. WCOT column (F) via a butt joint sealed by shrinkable Teflon tubing (G). The sample is injected by means of the syringe onto the bottom of the glass needle. After about a minute the flow of helium gas has evaporated most of the solvent through the vent. The glass needle is then pushed down into the oven (H) by means of the external magnet. The derivatized carbohydrates separate on the WCOT column and are detected by the FID detector (I). Typical gas pressures are: helium carrier gas 10 lb.in$^{-2}$, 4 ml/min through the vent, 0.5 ml/min through the column; nitrogen make-up gas 1 lb.in$^{-2}$, 1.0 ml/min; hydrogen detector gas 15 lb.in$^{-2}$; air detector gas 5 lb.in$^{-2}$.

27

**Table 11.** G.l.c. data of trimethylsilylated monosaccharides.

| Parent carbohydrate | Peak[a] | Time (min)[b] | Peak height[c] |
|---|---|---|---|
| (Solvent front) | 1 | 1.00 | |
| L-Arabinose | 2 | 4.82 | 0.271 |
| | 3 | 4.95 | 0.122 |
| | 6 | 5.36 | 0.024 |
| D-Ribose | 4 | 5.03 | 0.064 |
| | 5 | 5.26 | 0.317 |
| L-Rhamnose | 5 | 5.25 | 0.433 |
| | 6 | 5.36 | 0.046 |
| L-Fucose | 5 | 5.28 | 0.054 |
| | 7 | 5.52 | 0.266 |
| | 8 | 5.80 | 0.156 |
| D-Xylose | 9 | 6.10 | 0.377 |
| | 10 | 6.37 | 0.188 |
| D-Glucuronic acid | 11 | 7.14 | 0.069 |
| | 23 | 9.37 | 0.073 |
| | 24 | 9.49 | 0.240 |
| D-Galacturonic acid | 12 | 7.31 | 0.186 |
| | 13 | 7.90 | 0.056 |
| | 19 | 8.79 | 0.160 |
| | 20 | 8.90 | 0.051 |
| D-Mannose | 14 | 8.02 | 0.598 |
| | 16 | 8.42 | 0.042 |
| D-Galactose | 15 | 8.12 | 0.109 |
| | 17 | 8.54 | 0.382 |
| | 18 | 8.62 | 0.048 |
| | 21 | 8.97 | 0.153 |
| D-Glucose | 22 | 9.27 | 0.478 |
| | 25 | 9.59 | 0.201 |
| N-Acetyl-D-glucosamine | – | 10.57 | 0.031 |
| | – | 11.17 | 0.029 |
| | 28 | 11.70 | 0.359 |
| N-Acetyl-D-galactosamine | – | 10.45 | 0.031 |
| | 26 | 10.70 | 0.082 |
| | 27 | 11.35 | 0.257 |
| Myo-inositol | 29 | 12.24 | 1.000 |
| N-Acetylneuraminic acid | 30 | 15.35 | 0.200 |

[a]Refers to *Figure 7*.
[b]SD = 2 s.
[c]SD = 3.5%; peak height is proportional to peak area under these conditions.

(v) Vortex-mix the samples and evaporate to dryness using a stream of dry nitrogen (oxygen-free grade) at room temperature.

(vi) Add 50 μl of dry methanol, 5 μl of pyridine and 5 μl of acetic anhydride successively with intermediate mixing in order to re-N-acetylate the amino sugars (twice these volumes should be used if more than 50 nmol of amino sugars is present; fatty acid methyl esters may be removed, if suspected, by extraction with hexane).

(vii) Leave the solutions at room temperature for 15 min before evaporating to near

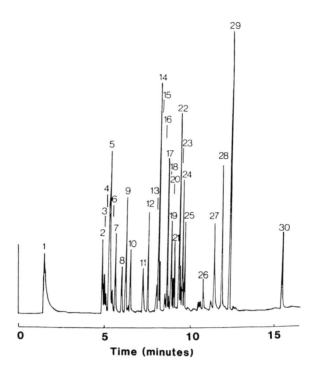

**Figure 7.** G.l.c. of trimethylsilylated monosaccharides (36). Approximately 0.5 nmol of each monosaccharide was derivatized (see Section 7.1). The peaks are identified in *Table 11*. Reprinted from ref. 36 with permission.

dryness by use of a stream of dry nitrogen (oxygen-free grade).

(viii)  Completely dry the residue using gentle vacuum over $P_2O_5$.

(ix)  After this thorough drying stage, add the silylation reagent (Reagent D, ~20−100 $\mu$l depending on the carbohydrate content), vortex-mix the tubes and leave for 1 h at room temperature.

The derivatized samples should be analysed the same day.

(x)  Re-evaporate using the dry nitrogen stream and immediately dissolve each sample in 14−50 $\mu$l of redistilled hexase.

(xi)  Apply some or all of this sample to the capillary column. The set-up recommended is a fused-silica WCOT column (25 m × 0.32 mm i.d.) coated with CP Sil 5 liquid phase (see *Table 9* for alternative liquid phases; any other reasonably equivalent fused silica column is suitable), using an all-glass solid injector [moving needle, Chrompack (*Figure 6*), previously inactivated by reaction with the silylation reagent Rejuv 8 (Supelco, but any similar product is suitable)]. Other injection techniques may be used but this method has the advantage that all the sample is injected onto the column without excessive solvent.

(xii)  Perform the analyses using dry oxygen-free helium as carrier gas and a temperature program (140°C for 2 min, then increasing at 8°C/min up to 240°C). A higher final temperature improves the analysis but tends to cause the more rapid deterioration of the shrinkable Teflon link between the injector and the column.

29

**Table 12.** Retention times and mass spectra primary fragments of methyl hexitol and hexitol acetates.

| Peak[a] | Positions of methylation[b] | Retention time[c] Silar 9CP | OV 101 | Primary fragments in the mass spectra |
|---|---|---|---|---|
| Fucitol | | —[d] | 1.107 | 73, 87, 145, 217, 289 |
| 1. | 2,3,4 | 0.808 | 0.864 | 117, 131, 161, 175 |
| Galactitol | | —[d] | 1385 | 73, 145, 217, 289 |
| 2. | 2,4 | 1.619 | 1.273 | 117, 189 |
| 3. | 3,6 | 1.512 | 1.208 | 45, 189, 233 |
| 4. | 4,6 | 1.439 | 1.201 | 45, 161, 261 |
| 5. | 2,3,4 | 1.431 | 1.176 | 117, 161, 189, 233 |
| 6. | 2,4,6 | 1.270 | 1.126 | 45, 117, 161, 233 |
| 7. | 3,4,6 | 1.327 | 1.126 | 45, 161, 189 |
| 8. | 2,3,4,6 | 1.086 | 1.022 | 45, 117, 161, 205 |
| Glucitol | | —[d] | 1.376 | 73, 145, 217, 289 |
| 32. | 3,4 | 1.593 | 1.197 | 189, 261 |
| 9. | 2,4,6 | 1.220 | 1.104 | 45, 117, 161, 233 |
| 10. | 3,4,6 | 1.236 | 1.099 | 45, 161, 189 |
| 11. | 2,3,4,6 | 0.991 | 0.997 | 45, 117, 161, 205 |
| Mannitol | | —[d] | 1.367 | 73, 145, 217 |
| 12. | 2 | 1.647 | 1.299 | 117 |
| 31. | 2,3 | 1.528 | 1.233 | 117 |
| 13. | 2,4 | 1.559 | 1.260 | 117, 189 |
| 14. | 2,6 | 1.415 | 1.180 | 45, 117 |
| 15. | 3,4 | 1.578 | 1.249 | 189 |
| 16. | 3,6 | 1.501 | 1.208 | 45, 189, 233 |
| 17. | 4,6 | 1.399 | 1.191 | 45, 161, 261 |
| 18. | 2,3,4 | 1.309 | 1.143 | 117, 161, 189, 233 |
| 19. | 2,3,6 | 1.290 | 1.109 | 45, 117, 161, 233 |
| 20. | 2,,4,6 | 1.230 | 1.118 | 45, 117, 161, 233 |
| 21. | 3,4,6 | 1.247 | 1.104 | 45, 161, 189 |
| 22. | 2,3,4,6 | 1.000 | 1.000 | 45, 117, 161, 205 |
| N-Methyl-2-amino-2-deoxy-galactitol | | —[d] | 1.553 | 158, 360 |
| 23. | 3,4,6 | 1.880 | 1.379 | 45, 158, 161, 202 |
| N-Methyl-2-amino-2-deoxy-glucitol | | —[d] | 1.529 | 158, 360 |
| 24. | 3 | —[e] | 1.520 | 202, 261, 346 |
| 25. | 6 | —[e] | 1.497 | 45, 158 |
| 26. | 3,4 | —[e] | 1.466 | 158, 189, 202 |
| 27. | 3,6 | 1.984 | 1.417 | 45, 158, 202 |
| 28. | 4,6 | —[e] | 1.462 | 45, 158, 161, 274 |
| 29. | 3,4,6 | 1.766 | 1.339 | 45, 158, 161, 202 |
| 30. | 1,3,5,6 | 1.439 | 1.191 | 45, 89, 116, 160, 205 |

[a]Refers to *Figure 8*.
[b]Numbers refer to positions of methylation. All the carbohdyrates are otherwise per-acetylated (e.g. mannitol 2,3,4,6 refers to 1,5-diacetyl-2,3,4,6-tetra-O-methyl-D-mannitol).
[c]Relative to 1,5-di-O-acetyl-2,3,4,6-tetra-O-methyl-D-mannitol (peak 22) = 1.000, on WCOT capillary columns (47).
[d]Not determined.
[e]Most partially methylated N-methylhexitol acetates are lost during chromatography over the polar Silar 9CP phase.

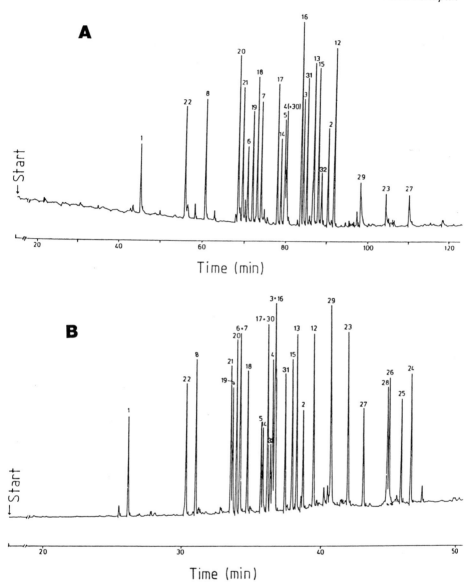

**Figure 8.** G.l.c. of peracetylated hexitols and methyl hexitols. 25 – 100 pmol of each of 32 *O*-methylated hexitol and 2-deoxy-2-(*N*-methyl)acetamidohexitol acetates, most of which are commonly found during the methylation analysis of *N*-glycosidically linked glycoprotein oligosaccharides, were analysed by (**A**) capillary g.l.c. using polar (22 m × 0.25 mm i.d. WCOT column containing Silar 9CP) and (**B**) nonpolar (85 m × 0.25 mm i.d. WCOT column containing OV 101) liquid phases. The peaks are identified in *Table 12*. The analytical conditions were (**A**) 20 min at 100°C then raised to 230°C at 1°C/min and (**B**) 100°C to 240°C at 3°C/min. Reprinted from ref. 47 with permission.

*Comments.* Most monosaccharides give a mixture of isomers on trimethylsilyl derivatization (*Table 11, Figure 7*). The ratio of the peak areas is constant, however, and may be used to confirm small amounts of material having a low signal-to-noise ratio. Labile carbohydrates (e.g. *N*-acetyl-neuraminic acid) are not destroyed by this analytical pro-

31

cess and it is therefore a good method for the analysis of all the carbohydrate moieties derived from glycoproteins. Fructose, however, is not analysable by this method and should be derivatized as the trimethylsilyl derivative of the free sugar, because the methyl glycoside is not formed on methanolysis. In general, the free sugar derivatives may be determined by g.l.c. analysis, as above, after trimethylsilylation of the dry material, without the methanolysis and re-acetylation steps. The retention times, number of peaks and relative peak heights will all be different to those obtained in the above method and should be determined separately. The D- and L-configurations of neutral monosaccharides may be determined in a similar manner but by the use of $R(-)$-2-butanol in place of methanol throughout. The trimethylsilylated $(-)$-2-butyl glycosides are separable on an SE30 capillary column (40).

### 6.2.2 D2: *Preparation of alditol acetate derivatives* (37)

*Reagents:*

(A)     Dry dimethyl sulphoxide, prepared by storage of material, from a previously unopened bottle, over molecular sieve type 4A. Care should be taken not to disturb any fine sediment that may be present when removing aliquots for use.

(B)     Dissolve 2 g of sodium borohydride in 100 ml of anhydrous dimethyl sulphoxide (reagent A) at 100°C. This reagent is stable if kept dry at 4°C.

*Method*

(i)     Dissolve the samples and standards containing up to about 500 $\mu$g of the monosaccharides in 0.1 ml of 1 M ammonia solution [prepared by diluting 3 ml of concentrated ammonia solution (sp. gr. 0.88) with 50 ml of water].

(ii)     Reduce the carbohydrates by the addition of 1 ml of reagent B followed by incubation at 40°C for 90 min.

(iii)     After reduction, decompose the excess sodium borohydride by the addition of 0.1 ml of glacial acetic acid.

(iv)     Per-acetylate the reduced monosaccharides by the addition of 0.2 ml of 1-methylimidazole followed by 2 ml of acetic anhydride and thorough mixing.

(v)     After 10 min at room temperature, add 5 ml of water to decompose the excess acetic anhydride.

(vi)     After cooling, extract the per-acetylated carbohydrates into the lower layer formed after vortex mixing the aqueous solution with 1 ml of dichloromethane. This layer may be removed via a Pasteur pipette and stored in a septum-capped vial at $-20$°C until analysis (*Table 12, Figure 8*).

*Comments.* Some carbohydrate pairs are reduced to the same alditol and cannot be resolved by this method, e.g. lyxose/arabinose, glucose/gulose and altrose/talose. In addition, ketoses are not reduced stereospecifically and give rise to isomeric hexitols, e.g. D-fructose gives D-glucitol and D-mannitol and D-sorbose gives L-glucitol and D-iditol.

### 6.2.3 *Inositol analysis*

The naturally occurring inositols can be simply separated by g.l.c. as either the hexakis-*O*-acetyl or hexakis-*O*-trimethylsilyl derivatives. The trimethylsilyl derivatives may be produced by the silylation reagent used for monosaccharide derivatization (Section 6.2.1)

**Table 13.** G.l.c. analysis of hexakis derivatives of naturally occurring inositols.

| Inositol isomer | Relative retention (myo-inositol = 1) | |
|---|---|---|
| | System 1[a] | System 2[b] |
| D-chiro- (1,2,4/3,5,6-) | 0.72[c] | 0.59 |
| L-chiro- (1,2,4/3,5,6-) | | 0.59 |
| neo- (1,2,3/4,5,6) | 0.81 | 0.47 |
| myo- (1,2,3,5/4,6, meso-, i-) | 1.00 | 1.00 |
| muco- (1,2,4,5/3,6-) | 1.17 | 0.51 |
| scyllo- (1,3,5/2,4,6-) | 1.53 | 0.79 |

[a]System 1: Hexakis-O-acetyl derivatives on a 2.5 m × 4 mm i.d. column [0.5% QF-1 plus 0.5% LAC-2R-466 on Chromosorb W HP (80−100 mesh) operated isothermally at 195°C (38)].
[b]System 2: Hexakis-O-trimethylsilyl derivatives on a 6 ft × 1/8 in i.d. column [3% SE 30 on silanized GAS-Chrom P (80−100 mesh) operated isothermally at 175°C (39)].
[c]It was not stated in ref. 38 whether the sample was the D- or L-isomer, or a mixture of the two.

but allowing the reaction to proceed for 48 h due to the slowness with which *neo*-inositol reacts. The acetyl derivatives may be produced by the 1-methylimidazole catalysed acetylation used for alditol derivatization (Section 6.2.2) or by suspending about 5 mg in 40 $\mu$l of glacial acetic acid, adding 100 $\mu$l of trifluoroaceteic anhydride and warming gently to initiate the reaction. After 10 min at room temperature the reaction mixture should be rotary-evaporated and the residue dissolved in methylene chloride for g.l.c. analysis. The g.l.c. analyses of mixtures of both these derivatives is shown in *Table 13*.

## 7. MASS SPECTROMETRY

This extremely powerful analytical technique is usually used in conjunction with g.l.c. (as g.l.c.-m.s.). Together they have become the fastest technique for the analysis of complex mixtures using two sets of data, the retention times and the mass spectra. M.s. is based on the positive ionization of molecules due to bombardment with a beam of electrons. As carbohydrates are normally converted into volatile derivatives before analysis, the linkage of m.s. to g.l.c., which also requires volatile derivatives, is clearly beneficial. The mass spectra from monosaccharide derivatives consist of primary fragments formed from the initial cleavage of the molecular ion plus many secondary fragments due to the elimination of neutral molecules from these primary fragments. The fragmentation pattern may be used for identification by comparison with spectra from known materials or by deduction from the known cleavage and elimination probabilities. In general, the molecular ion of a derivatized carbohydrate is not seen and large primary fragments tend to eliminate neutral molecules quite readily. Most of the usable spectra are , therefore, below 300 m/e. Care should also be taken not to over-rely on the absolute or relative peak heights in the spectra as these may vary considerably from day to day, depending on the precise conditions used. However if care is taken, and suitable peaks are chosen, a reproducibility of better than ±10% is possible. One further drawback to the use of mass spectrometry is that there is generally little difference between isomers. This is easily overcome by the additional use of the retention data. G.l.c.−m.s. is a particularly powerful technique for the analysis of partially methylated alditol acetates (*Figure 8, Table 12*) especially as known standards are not

```
                              HCOAc              HC=O
                               ‖                  |
                               HC  ───────→      HCH
                            +  |               +  |
                            MeO=CH             MeO=CH
                            ↗ 129                 87
        , CH₃-C=O     H
          |  ‖        HCOAc           HCH            HCH
       H  ¦  43 ,       |              ‖              |
      HCOAc     ,     HCOAc  ───→    HCOAc   →       COAc
        |            +  |          +  |              ‖
      HCOAc          MeO=CH        MeO=CH           HCH
        |              189           129             99
      MeOCH
    ─ ─ | ─ ─ ─ ─ ─ ─ HC=OMe
      HCOMe     161    |
        |              HCOAc
      HCOAc            |               HC=OMe
    ─ ─| ─ ─ ─ ─ ─    HCOMe           |
      HCOMe     45     H              HC
        H                             ‖
       CH₂=OMe                        HCOMe
                                      101
```

**Figure 9.** The fragmentation pattern in mass spectrometry of 1,2,5-triacetyl-3,4-6-tri-*O*-methyl-D-glucitol.

```
      |                 |                 |
      HC-OMe            HC-OMe            HC-OAc
    ··|······   >     ··|······   »     ··|······
      HC-OMe            HC-OAc            HC-OAc
      |                 |                 |
                  ⫽          ⬊
               |  H               |  CH₃
               HC-NAc            HC-NAc
             ··|······         ··|······
               HC-OAc            HC-OAc
               |                 |
```

**Figure 10.** The relative ease of cleavage, in mass spectrometry between carbon atoms in peracetylated methylated alditols and amino-alditols.

always available. The mass spectra may be analysed with reference to a few simple rules (see also *Figure 9*);

(i)  Fission prefers to take place next to the methoxy or *N*-acetyl groups rather than the *O*-acetyl groups, with the positive ion stabilized by the methoxyl or *N*-acetyl grouping (*Figure 10*).

(ii)  Secondary fragments are derived from the primary fragment by the single or consecutive elimination of formaldehyde (mol. wt 30), methanol (mol. wt 32), ketene (mol. wt 42) or acetic acid (mol. wt 60), (e.g. R-CH=Ṅ(CH₃)Ac fragments immediately eliminate ketene to form R-CH=ṄHCH₃).

(iii)  Large primary fragments eliminate neutral molecules more readily than small primary fragments. A base fragment at m/e 43 (CH₃Ċ=O) is normally found.

There is no significant difference between isomeric alditols having the same substitution pattern, alditols from 2,3- and 3,4-di-*O*-methyl pentoses or from 3- and 4-*O*-methyl-hexoses would be indistinguishable unless sodium borodeuteride was used in the reduction step.

The trimethylsilyl derivatives of alditols, monosaccharides or methyl glycosides cleave between carbon atoms carrying trimethyl silyl ether or methoxyl groups. The secondary fragments are formed from these primary fragments by the consecutive elimination of trimethylsilylhydroxide (mol. wt 90) molecules. A base fragment at m/e 73 $[(CH_3)_3Si^+]$ is normally found. The mass spectra of the trimethylsilyl derivatives of cyclic carbohydrates are complicated because the initial cleavage does not reduce the m/e of the resulting fragment; secondary cleavages must occur and these produce fragments at m/e 147 ($Me_3Si\text{-}\overset{+}{O}=Si\ Me_2$, from pairs of derivatized hydroxyl groups) 204 $[(Me_3SiO\text{-}CH_2)_2{}^+$, from vicinal hydroxyl groups] and 217 ($Me_3SiO\text{-}CH=CH\text{-}\overset{+}{C}H\text{-}OSiMe_3$ from three neighbouring hydroxyl groups) in addition to the base peak at m/e 73 and a peak at m/e 191 $[(Me_3SiO)_2\overset{+}{C}H]$ due to the straight chain form at C-1. The ratio of these 5 peaks may be used to distinguish between carbohydrates if reproducible analytical conditions are possible. In particular, for aldohexoses, aldopentoses and 6-deoxy-aldohexoses, the pyranoid forms give a ratio m/e 204/217 $\geq 1$ ($\sim 1$ to $> 1$ for aldopentoses), whereas the furanoid forms show a ratio $\leq 1$. An intense peak at m/e 205 ($Me_3SiOCH_2\text{-}CH=\overset{+}{O}SiMe_3$) is also characteristic of a furanoid ring in aldohexoses.

## 8. NUCLEAR MAGNETIC RESONANCE

Both $^1$H-n.m.r. and $^{13}$C-n.m.r. spectroscopy are relevant to monosaccharide analysis. $^1$H-n.m.r. is the more sensitive technique, $^{13}$C-n.m.r. being only 1.6% as sensitive and depending on the natural abundance of $^{13}$C (1.1%) for its application. They are closely related techniques but give sufficiently different data that they complement each other. Often, however, $^1$H-n.m.r. spectra of monosaccharides are poorly resolved presenting little analysable information and are only useful for identification by comparison with known spectra. Their main advantage lies in requiring relatively small amounts of sample ($\sim 10\ \mu g$, 0.9 ml of 0.05 mM in $D_2O$). Although substantially larger samples are required, $^{13}$C-n.m.r. spectra are more useful, particularly for the determination of the structure of unknown samples. Proton-decoupled $^{13}$C-n.m.r. spectra are well resolved and usually provide an unambiguous identification of a compound, especially when compared with a collection of $^{13}$C-n.m.r. chemical shifts (e.g. ref. 48). The normal solvents are $D_2O$, perdeuterated dimethyl sulphoxide and $CDCl_3$, of which $D_2O$ is the preferred solvent for most monosaccharides and their derivatives although it is highly hygroscopic. The choice of solvent or concentration of samples have little effect on the decoupled $^{13}$C-n.m.r. spectra of neutral monosaccharides although the spectra of charged monosaccharides is pH dependent. Samples for analysis should be as pure as possible and free from insoluble or paramagnetic material. Dissolved oxygen should be removed by boiling for a very brief period ($\sim 1$ min).

## 9. INFRA-RED SPECTROSCOPY

Infra-red spectrscopy is mainly of use in monosaccharide analysis for the confirmation of the identity of a molecule by comparison with a published standard spectrum or the spectra of known material. About 1 mg of pure dry carbohydrate is needed for an analysis although a smaller amount may give useful spectra. The preferred method makes use of salt discs made by the intimate mixture of about 1 mg of carbohydrate with 300 mg of pure dry KBr followed by pressing into a disc. The whole i.r. spectrum ($4000-650$ cm$^{-1}$) should be used for comparison (50).

# 10. REFERENCES

1*. Dische,Z., Shettles,L.B. and Osnos,M. (1949) *Arch. Biochem.*, **22**, 169.
2*. Dubois,M., Gilles,K.A., Hamilton,J.K., Rebers,P.A. and Smith,F. (1956) *Anal. Chem.*, **28**, 350.
3*. Dygert,S., Li,L.H., Florida,D. and Thoma,J.A. (1965) *Anal. Biochem.*, **13**, 367.
4*. White,C.A.and Kennedy,J.F. (1981) Techniques in the life sciences *B3*, B312/1.
5*. Bernfeld,P. (1955) in *Methods in Enzymology*, Colowick,S.P. and Kaplan,N.O. (eds.), Academic Press, New York, Vol. **1**, p. 149.
6. Boratynski,J. (1984) *Anal. Biochem.*, **137**, 528.
7*. Reissig,J.L., Strominger,J.L. and Leloir,L.F. (1955) *J. Biol. Chem.*, **217**, 959.
8. Bitter,T. and Muir,H.M. (1962) *Anal. Biochem.*, **4**, 330.
9*. Warren,L. (1959) J. Biol. Chem., **234**, 1971.
10. Svennerholm,L. (1957) *Biochim. Biophys. Acta*, **24**, 604.
11*. Bergmeyer,H.U., Bernt,E., Schmidt,F. and Stork,H. (1974) in *Methods of Enzymatic Analysis*, Bergmeyer,H.U. (ed.), Verlag Chemie, Academic Press, New York and London, Vol. **3**, p. 1196.
12*. Anon (1984) in *Methods of Enzymatic Food Analysis*, Boehringer Mannheim GmbH, Mannheim, p. 65.
13*. Anon (1984) in *Methods of Enzymatic Food Analysis*, Boehringer Mannheim GmbH, Mannheim, p. 71.
14*. Morris,J.B. (1982) *Anal. Biochem.*, **121**, 129.
15. Fujimura,Y., Ishii,S., Kawamura,M. and Naruse,H. (1981) *Anal. Biochem.*, **117**, 187.
16. Sugiura,M., Hayakawa,S., Ito,Y. and Hirano,K. (1981) *Chem. Pharm. Bull.*, **29**, 146.
17. Ghebregzabher,M., Rufini,S., Monaldi,B. and Lato,M. (1976) Chromatographic Reviews 95. *J. Chromatogr.*, **127**, 133.
18. Ghebregzabher,M., Rufini,S., Sapia,G.M. and Lato,M. (1979) *J. Chromatogr.*, **180**, 1.
19. Zweig,G. and Sherma,J. (eds.) (1972) *CRC Handbook of Chromatography, Vol. 1*, and Churms,S.C. (ed.) (1982) *Vol. I, Carbohydrates*, CRC Press Inc., Boca Raton, Florida.
20. Sherma,J. and Zweig,G. (1971) *Paper Chromatography and Electrophoresis. Vol. 2. Paper Chromatography*, Academic Press, New York and London.
21. Mes,J. and Kamn,J. (1968) *J. Chromatogr.*, **38**, 120.
22. Robyt,J.F. (1975) *Carbohydr. Res.*, **40**, 373.
23. Hough,L. and Jones,J.K.N. (1962) in *Methods in Carbohydrate Chemistry*, Whistler,R.L. and Wolfrom,M.L. (eds.), Academic Press, New York and London, Vol. **1**, p. 21.
24. Hansen,S.A. (1975) *J. Chromatogr.*, **107**, 224.
25. Got,R., Cheftel,R.-I., Font,J. and Moretti,J. (1967) *Biochim. Biophys. Acta*, **136**, 320.
26. Grimble,G.K., Barker,H.M. and Taylor,R.H. (1983) *Anal. Biochem.*, **128**, 422.
27. Honda,S., Takahashi,M., Shimada,S., Kakehi,K. and Ganno,S. (1983) *Anal. Biochem.*, **128**, 429.
28. Honda,S., Takahashi,M., Kakehi,K. and Ganno,S. (1981) *Anal. Biochem.*, **113**, 130.
29. Honda,S. and Suzuki,S. (1984) *Anal. Biochem.*, **142**, 167.
30. White,C.A., Kennedy,J.F. and Golding,B.T. (1979) *Carbohydr. Res.*, **76**, 1.
31. Takemoto,H., Hase,S. and Ikenaka,T. (1985) *Anal. Biochem.*, **145**, 245.
32. Alpenfels,W.F. (1981) *Anal. Biochem.*, **114**, 153.
33. Dethy,J.-M., Callaert-Deveen,B., Janssens,M. and Lenaers,A. (1984) *Anal. Biochem.*, **143**, 119.
34. Golik,J., Liu,H.-W., Dinovi,M., Furukawa,J. and Nakanishi,K. (1983) *Carbohydr. Res.*, **118**, 135.
35. Nordin,P. (1983) *Anal. Biochem.*, **131**, 492.
36. Chaplin,M.F. (1982) *Anal. Biochem.*, **123**, 336.
37. Blakeney,A.B., Harris,P.J., Henry,R.J. and Stone,B.A. (1983) *Carbohydr. Res.*, **113**, 291.
38. Irving,G.C.J. (1981) *J. Chromatogr.*, **205**, 460.
39. Wells,W.W., Pittman,T.A. and Wells,H.J. (1965) *Anal. Biochem.*, **10**, 450.
40. Gerwig,G.J., Kamerling,J.P. and Vleignethart,F.G. (1978) *Carbohydr. Res.*, **62**, 349.
41. Kennedy,J.F. and Fox,J.E. (1977) *Carbohydrate Res.*, **54**, 13.
42. Shukla,A.K., Scholz,N., Reimerdes,E.H. and Schauer,R. (1982) *Anal. Biochem.*, **123**, 78.
43. Perini,F. and Peters,B.P. (1982) *Anal. Biochem.*, **123**, 357.
44. Krantz,M.J. and Lee,Y.C. (1975) *Anal. Biochem.*, **63**, 464.
45. Honda,S. (1984) *Anal. Biochem.*, **140**, 1.
46. Sutton,G.S. (1973) *Adv. Carbohydr. Chem. Biochem.*, **28**, 11.
47. Geyer,R., Geyer,H., Kühnhardt,S., Mink,W. and Stirm,S. (1982) *Anal. Biochem.*, **121**, 263.
48. Bock,K. and Pedersen,C. (1983) *Adv. Carbohydr. Chem. Biochem.*, **41**, 27.
49. Chaplin,M.F. (1985) Unpublished results.
50. Tipson,R.S. and Parker,F.S. (1980) in *The Carbohydrates*, 2nd Ed., Vol. **1B**, Pigman,W.W. and Horton, D. (eds.), Academic Press, New York, pp. 1394.

*Indicates a modification of the cited method.

# CHAPTER 2

# Oligosaccharides

CHARLES A.WHITE AND JOHN F.KENNEDY

## 1. INTRODUCTION

Oligosaccharides are traditionally defined as polymers of monosaccharides containing from two to 10 residues. However, since the naturally occurring polysaccharides rarely contain less than 25−30 residues, it is possible to consider the range of polymers having between two and about 20−25 residues as oligosaccharides, particularly when considering the analytical methods involved. As oligosaccharides fall into the classification between monosaccharides and polysaccharides it is not surprising that the methods used for oligosaccharide analysis are extensions of those used for either monosaccharide or polysaccharide analysis. This chapter uses a slightly different format to that of Chapter 1 in order to give the maximum information, within the limitations of available space, on the modifications which have to be made in adapting the methods described in detail for the analysis of monosaccharides and polysaccharides. Rather than repeat descriptions of practical details etc. this chapter concentrates on the types of methods which can be applied and their relative merits and shortcomings.

As with monosaccharide analysis, the analysis of oligosaccharides can be divided into qualitative identification and quantitative determination but, in addition to determination of the number and type of component residues, the actual linkage between successive residues is an area not encountered in monosaccharide analysis. Due to the complexities of linkage analysis, discussion of the methodology is not described in this chapter but can be found in subsequent chapters concerning the analysis of polysaccharides and carbohydrate-containing macromolecules. The traditional methods used for qualitative analysis: paper chromatography and thin layer chromatography (t.l.c.), have largely been replaced by rapid quantitative techniques, but, for completeness, are described briefly. Methods for the separation and quantitation of mixtures of oligosaccharides and for the identification and quantitation of individual purified oligosaccharides are described. Rather than providing a listing of all methods which have been devised, only those which have become the more commonly used techniques are described.

## 2. COLORIMETRIC METHODS

The colorimetric methods which have been devised are mainly used for the gross determination of total carbohydrate content or total reducing sugar content although specific assay methods have been developed for the quantitation of an individual oligosaccharide from a mixture of compounds via the use of specific enzymes. Both total carbohydrate and reducing sugar contents are important features of oligosaccharide analysis since it is possible to obtain a value for the size of an oligosaccharide or, for a mixture of related oligosaccharides, a mean value for the ratio of total residues to terminal reducing residues. With the development of sugar syrups (1) methods for the characterization of these products have been developed with the various syrups being categorized by their dextrose (i.e. D-glucose) equivalent or DE value. Despite the shortcomings of defining mixtures of oligosaccharides in terms of this one parameter (2) it is still used as the principle method for characterizing oligosaccharide mixtures.

## 2.1 Total sugar assays

A number of assays have been developed which rely on the action of concentrated (or near concentrated) sulphuric acid causing hydrolysis of all glycosidic linkages and the subsequent dehydration of the monosaccharides released to give derivatives of furfural (e.g. hexoses produce 5-hydroxymethyl furfuraldehyde). The dehydration products react with a number of compounds such as L-cysteine (3), phenol (4), orcinol (5) and anthrone (6) to give coloured products. Whilst total sugar concentrations are readily obtained for homoglycans, care must be taken in interpreting the results obtained with heteroglycans due to the different colour intensities produced by different monosaccharides. Practical details for the performance of the first two assays mentioned can be found in Chapter 1.

### 2.2.1 *Orcinol-sulphuric acid assay*

*Sensitivity:* $0-20$ $\mu$g carbohydrate in 200 $\mu$l
*Final volume:* 1.0 ml
*Reagent:* Ice-cold orcinol (recrystallized from benzene) dissolved in concentrated acid (2 g/l). This reagent should be prepared fresh each day but can be stored at 4°C for up to one week.

*Method:*

(i)     To precooled (to 4°C) samples, standards and controls (200 $\mu$l) add 800 $\mu$l of reagent with care. Mix well.
(ii)    Heat the solutions at 80°C for 15 min and cool rapidly to room temperature.
(iii)   Determine the absorbance at 420 nm.

*Comments:* The original method (4) suggested that absorbances should be determined at 510 nm but the use of 420 nm results in an increased sensitivity (by a factor of two) and a reduced interference from uronic acids and deoxy sugars.

## 2.2 Reducing sugar assays

Traditionally the Lane and Eynon (7) assay was used to determine the content of reducing sugars in a sample and still is the method of choice in some industrial applications. The method involves the reaction of reducing sugars with alkaline cupric salts to give curpous oxide which can be monitored titrimetrically to give the concentration of various reducing sugars by reference to standard tables. Recent modifications to this assay have eliminated the use of tables (8).

A far more convenient assay is that which involves the reaction with alkaline 3,5-dinitrosalicylic acid originally devised by Bernfeld (9) and for which practical details are given in Chapter 1. The major disadvantages of this assay are its insensitivity at low carbohydrate concentrations (10) and some workers report different responses obtained for equimolar amounts of oligosaccharides (11). For this reason, the less convenient Nelson-Somogyi (12) assay is preferred in some instances (13). This assay is based on the reduction of cupric salts to cuprous salts which further reduce the arsenomolybdate complex to molybdenum blue, the intensity of which is determined spectrophotometrically. It is however less sensitive than the 3,5-dinitrosalicyclic acid assay.

A number of other assays have been reported such as those using alkaline ferricyanide (14) or alkaline picric acid (15). A recent innovation is the reaction based on the production of coloured soluble formazan salt from the dye tetrazolium blue (16) which, as a result of an essential solvent extraction stage, can be applied to samples containing particulate matter.

### 2.2.1 *Tetrazolium Blue assay*

*Sensitivity:* $0-10$ $\mu$g glucose equivalent in 100 $\mu$l
*Final volume:* 1.0 ml
*Reagent:* Add three volumes of 0.3 M NaOH to one volume of an aqueous suspension of Tetrazolium Blue (1% w/v) and stir until completely dissolved. Dilute with five volumes distilled water and store at 5°C in the dark.

*Method:*

(i)     To samples, standards and controls (100 $\mu$l) add 900 $\mu$l of reagent and mix well.
(ii)    Heat the solutions at 100°C for 30 sec and cool rapidly to room temperature.
(iii)   Add 1 ml of toluene and shake until no more colour is extracted from the aqueous layer.
(iv)    Determine the absorbance at 570 nm.

*Comments:* The sensitivity of the assay can be adjusted to accommodate small or large colour changes by halving or doubling the volume of toluene used for the extraction stage. Extended heating of reaction mixture results in alkaline hydrolysis of oligosaccharides and increased absorbance values.

### 2.3 **Automated assay systems**

One of the major uses of colorimetric assays is in the monitoring of chromatographic columns. Traditionally this was done by collecting fractions of the effluent stream and performing series of manual assays. To a large extent this has now been replaced by either non-specific detectors (see Section 5.6) or automated assay systems of various degrees of specificity. The pioneering work on assay automation was performed using air segmented systems based on glass tubing of approximately $1.6-2.4$ mm i.d. and Technicon AA1 type equipment. This allows mixing of reagents via peristaltic pumps equipped with tubing of different sizes and construction to facilitate accurate mixing of the majority of reagents including certain organic solvents and concentrated mineral acids. Increasingly the assay systems are now being devised use non-segmented flow-through narrow bore (0.5 mm i.d. and smaller) Teflon tubing but such systems require pumps capable of pumping solvents and corrosive liquids at the higher pressures resulting from the use of the narrow bore tubing.

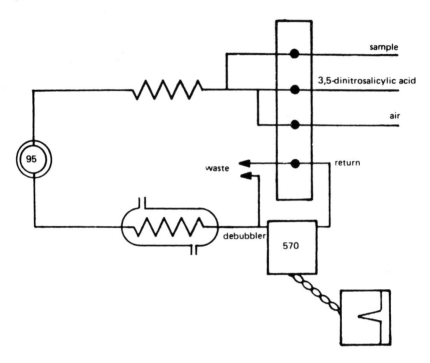

**Figure 1.** The automated 3,5-dinitrosalicylic acid assay. (DNS reagent consisting of 1.0 g of 3,5-dinitrosalicylic acid and 300 g of potassium sodium tartrate, per litre of 0.4 M NaOH 0.73 ml/min; sample, 0–3 mg D-glucose equivalent/ml water or buffer, 0.20 ml/min; air, 0.73 ml/min; waste, 0.73 ml/min).

Several systems have been devised for the detection of reducing sugars (see Chapter 1) but these are only applicable to the detection of the smaller oligosaccharides due to the response being related to the molar concentration of the oligosaccharide and not the concentration by weight. Since the response of maltohexaose is only 18% of that obtained for the same weight of D-glucose, reducing sugar assays are never used for the detection of oligosaccharides above hexasaccharides and ideally are best suited to the detection of disaccharides and trisaccharides. The automated 3,5-dinitrosalicylic acid assay is shown in *Figure 1* by way of an example.

The detection of total sugar content relies on the use of reagents such as concentrated acid to hydrolyse the oligosaccharides to monosaccharides and to produce a suitable chromogen (e.g. 5-hydroxymethyl furfuraldehyde) and consequently the development in suitable automated assay systems has not received the same degree of attention. However systems have been devised for the estimation of total sugar content based on the L-cysteine (17) (see *Figure 2*) or orcinol assays (18) for neutral oligosaccharides and the carbazole (17) assay for uronic acid-containing oligosaccharides using segmented (17) or non-segmented (18) flow systems. The combination of such systems and chromatographic columns has led to the development of fully automated oligosaccharide analysers based on ion exchange (18) or gel permeation (19) chromatography.

## 3. THIN LAYER CHROMATOGRAPHY

T.l.c. is not widely used for oligosaccharide analysis, but can be useful as a rapid techni-

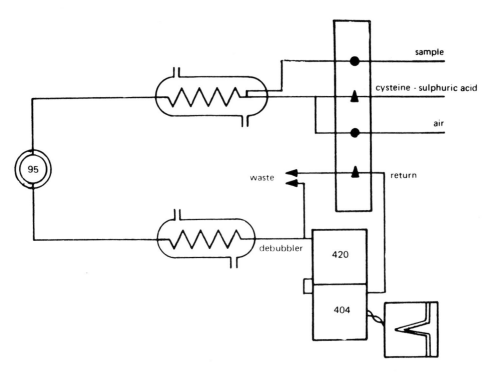

**Figure 2.** The automated L-cysteine-sulphuric acid assay. (L-cysteine hydrochloride reagent, 700 mg/ml 86% sulphuric acid, 0.53 ml/min; sample, $0-50$ $\mu$g/ml water or buffer, 0.1 ml/min; air, 0.23 ml/min; waste 0.32 ml/min).

que for monitoring, for example, the hydrolysis of starch to give a particular oligo-saccharide spectrum. It is also useful when it is not possible to perform more complex studies, as for example during field studies. The technique is essentially qualitative in its execution although semi-quantitative data can be obtained using scanning densitometric analysis of the visualized chromatogram. Such results must however be interpreted with care since the degree of colour produced by the various visualization techniques is affected by parameters such as temperature and duration of heating, coverage of spray reagent, traces of salts or eluants, etc. all of which can vary on a single chromatogram as well as between different chromatograms.

Two systems which can be used to provide initial data on the relative abundance of the lower members of homologous series of oligosaccharides are described briefly. Both use the multiple-ascents technique, whereby the chromatogram is eluted in the normal manner, dried and re-eluted with the same eluant a number of times, in order to improve the resolution between components. Full practical details for t.l.c. analysis can be found in Chapter 1.

Kieselgel G has been used to separate the individual members of a series of non-reducing oligofructosides, up to a hexasaccharide, derived from plant extracts by applying the mixture in 70% ethanol and eluting the chromatogram with chloroform-acetic acid-water (3:3.5:0.5) at ambient temperature with three ascents being used for optimum resolution (20). The individual members of the series of oligosaccharides derived from the enzymic hydrolysis of starch, up to maltodecaose, can be separated using

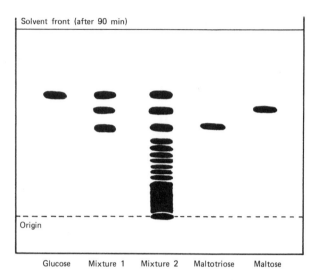

**Figure 3.** T.l.c. separation of standard malto-oligosaccharides on silica gel G using butanol-ethanol-water (5:3:2) eluant. (Individual components, 4 mg/ml, 5 μl loaded; mixture 1 = D-glucose, maltose and maltotriose, 2 mg each/ml, 10 μl loaded; mixture 2 = starch hydrolysate 50 mg/ml, 5 μl loaded).

silica gel G and an eluant of butanol-ethanol-water (5:3:2) with three ascents for optimum resolution (21) (see *Figure 3*).

Detection of components on a thin layer chromatogram can readily be achieved using a non-selective charring technique by heating the chromatogram after spraying it with a reagent prepared from concentrated sulphuric acid in ethanol (1:9). This method can be improved such that increased contrast and sensitivity is obtained if ceric sulphate (3% w/v) is added to the sulphuric acid-ethanol spray (21). Selective detection can be obtained using spray reagents containing diphenylamine aridine phosphate (22) or *p*-anisaldehyde (23) which give different colours with different carbohydrates and thereby assist in identifying overlapping or unresolved peaks. The use of reagents such as tetrazolium blue and its derivatives (see Chapter 1), which only visualize reducing sugars, are of little value for oligosaccharide analysis due to the diminishing response with increasing molecular weight of oligosaccharides.

## 4. LOW PRESSURE COLUMN CHROMATOGRAPHY

The distinction between low pressure (traditional) column chromatography and high performance liquid chromatography (h.p.l.c.) is not as clear cut as many workers would have you believe, with definitions being based on particle size (25 μm is a typical division) or the capital cost of equipment. Consequently some of the discussion in this section can equally apply to Section 5 and vice versa. Low pressure column chromatography is typified by the use of compressible column packings which require the use of low pumping pressures (hydrostatic pressure or peristaltic pumps frequently being used) and extended analysis times in the region of 2−18 h. Detection is frequently performed using automated assay systems (Section 2.3) or by manual assay following collection of eluant fractions. Many of the historical developments and applications with respect to oligosaccharide analysis have been reviewed recently (24).

## 4.1 Ion-exchange chromatography

### 4.1.1 *Types of systems available*

Ion-exchange chromatography has almost universally replaced the traditional adsorption chromatography based on charcoal-celite mixtures developed in the early 1950s which was capable of separating series of oligosaccharides up to a degree of polymerization (d.p.) of 8 − 10 using ethanol or butanol gradients in water as eluant (25). The use of ion-exchange resins is preferred due to their increased selectivity, lower operational back pressures, increased reproducibility and the ability to use eluants which do not decrease the solubility of the higher oligosaccharides as is the case with mixtures of alcohols in water.

Anion-exchange resins, most commonly in the carbonate or bicarbonate form, although chloride and hydroxyl forms can be used, will fractionate oligosaccharides in order of decreasing molecular size. The use of water as eluant allows fractionation of neutral sugars up to d.p. about 6 but the possibility of interconversion of terminal residues under the effects of the basic ion-exchange resins has prevented a full exploitation of the method. The use of acetic acid, sodium acetate, formic acid or lithium chloride eluants allows the fractionation of acid-containing oligosaccharides such as uronic acid- or aldonic acid-containing oligosaccharides with, for example, up to d.p. 14 being fractionated for the series of xylonic acids prepared from a birch xylan hydrolysate using Dowex 1-X2 (26).

Cation-exchange resins have also been used for separation of oligosaccharides in an attempt to overcome the effects of sugar interconversions by anion-exchange resins. Cation-exchange resins in the lithium, barium or potassium form have been used to give results similar to those obtained with anion-exchange resins but with the aid of less complex eluants. The effect of the counterion is considerable and the correct choice is essential for optimum resolution of a given oligosaccharide mixture (27).

Use of water-alcohol mixtures to elute ion-exchange columns results in a partition mechanism being responsible for the separation, with equilibria being established between the phase within the resin and the bulk solution. With anion-exchange resins there is a higher proportion of water within the resin than in the bulk solution which favours retention of the higher oligosaccharides such that a reversal of elution order can be achieved compared to anion-exchange chromatography using water as sole eluant, with monosaccharides eluting before disaccharides and the higher oligosaccharides following. This results in incomplete recovery of some mixtures of higher oligosaccharides and restriction of the method to mixtures containing d.p. less than 10. The most common resin used is the sulphate form of styrene-divinylbenzene resins (28) although quaternary ammonium salts derived from polymeric carbohydrate resins (i.e. cross-linked dextran) can be used (29). The use of cationic resins in this partition-type mode is of little use in oligosaccharide analysis due to the high alcohol concentrations required for elution. This restricts the analysis to disaccharides only.

If borate buffers are used to form complexes with carbohydrates the different affinities of the negatively charged complexes for the ion-exchange resins can be exploited as a means of separating oligosaccharides as an extension to the normal ion-exchange chromatographic separation of monosaccharides (18) (see *Figure 4*). The extended analysis times required and the alkaline conditions can cause interconversion of the terminal residues of reducing sugars. By alteration of the composition of the eluting

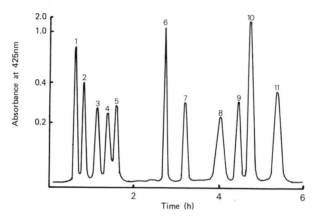

**Figure 4.** Ion-exchange chromatography of monosaccharides and disaccharides using Jeol LC-R-3 ion exchange resin (1, sucrose; 2, cellobiose; 3, maltose; 4, lactose; 5, rhamnose; 6, ribose; 7, mannose; 8, fucose; 9, galactose; 10, xylose; 11, glucose).

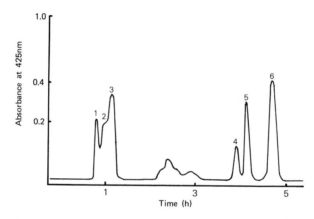

**Figure 5.** Ion-exchange chromatography of starch-derived oligosaccharides showing structural features enhanced by borate ion complexation using Jeol LC-R-3 resin (1, panose; 2, maltotriose; 3, maltose; 4, isopanose; 5, isomaltose; 6, D-glucose).

buffers the system is ideally suited for separation of families of structurally related disaccharides and trisaccharides such that structural features of, for example, reversion products of enzymically degraded starch can be identified (30) (see *Figure 5*).

### 4.1.2 *Experimental details*

The chosen ion-exchange resin is converted to the desired form and solvated by the methods described by the manufacturers of the resins. A glass column (diameter in the range of 0.6−1.0 cm, and length 10−30 cm) which has been drawn to a neck at the lower end and which contains a small plug of glass wool is filled with buffer and the resin poured in and allowed to settle. The use of proprietary columns and fittings such as the Omnifit® range (Omnifit Ltd) greatly assists in the production of a chromatographic column which can be pumped under increased pressure or which flows under hydrostatic pressure. The resin is pumped with buffer or eluant until a stable

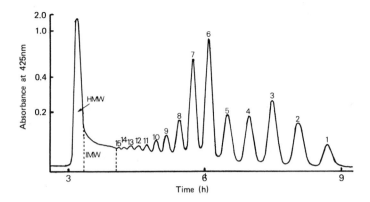

**Figure 6.** Gel permeation chromatographic fractionation of starch-derived oligosaccharides on Bio-Gel P2. (1,2,3 etc., D-glucose, maltose + isomaltose, D-glucotrisaccharides, etc., IMW, intermediate molecular weight material d.p. 15 to approximately 20; HMW, high molecular weight above d.p. approximately 20).

packed bed is obtained.

The sample is introduced onto the column by use of a proprietary system injector. Alternatively it can be introduced by removal of the eluant head from above the resin, allowing the sample to enter the resin (with the aid of air pressure from a syringe attached via a rubber bung to the top of the column if necessary), washing the sample into the resin with a small amount of eluant and replacing the eluant head.

The column is pumped with eluant and monitored either by automated assay (Section 2.3) or by collection of fractions and manual assay (Section 2.1).

## 4.2 Gel permeation chromatography

Gel permeation chromatography (g.p.c.) is defined as the separation of compounds according to their molecular weight, or more correctly their hydrodynamic volume. It has been developed into a very useful technique for oligosaccharide analysis. Tridimensional cross-linked gels, based on dextran (e.g. Sephadex G10 or G25 from Pharmacia) or polyacrylamide (e.g. Bio-Gel P2 from Bio-Rad Laboratories) have the degree of cross-linking controlled such that the resultant pores are too small to allow high molecular weight species to penetrate. They are excluded from the gel and therefore elute in a volume equal to the void (dead) volume ($V_o$) of the column. Very small molecules penetrate the gel freely and are eluted in a volume ($V_T$) equal to the sum of the void volume and the volume of solvent within the gel matrix available to small molecules (i.e. the pore volume, $V_p$). Thus

$$V_T = V_o + V_p$$

Molecules of intermediate size can penetrate some of the pores of the gel matrix and the degree of penetration, which is related to the distribution of pore sizes and the hydrodynamic volume of these intermediate size molecules is reflected in the elution volume ($V_e$) which lies within the range $V_o$ and $V_T$ (see *Figure 6*). The distribution of pore sizes within the gel matrix is such that over a considerable range of elution volumes the relationship:-

$$V_e \propto \log \text{(molecular weight)}$$

can be applied. At the extremities of the range of elution volume this relationship does

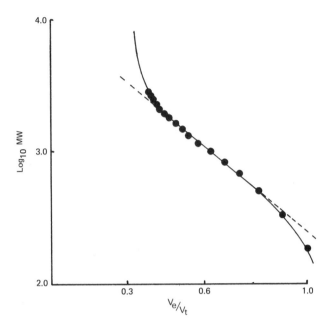

**Figure 7.** Calibration curve for Bio-Gel P2 derived from the cello-oligosaccharide series obtained by the controlled hydrolysis of cellulose.

not apply, but it is still possible to obtain some data provided a calibration curve (see *Figure 7*) is constructed to cover the range of elution volume involved, using materials having similar known hydrodynamic volumes. Examples have been published (31,32) which show that homologous series of oligosaccharides have different gradients for their calibration curves due to the different hydrodynamic volumes for oligosaccharides having the same molecular weight. Therefore it is important to exercise care in interpreting the molecular weight of an unknown oligosaccharide when using a gel permeation column which has been calibrated with, for example, the malto-oligosaccharide series (derived from the controlled hydrolysis of amylose) or the cello-oligosaccharide series (as shown in *Figure 7*).

In an ideal system the only effects which contribute to the separation of oligosaccharides are steric effects but it has now become well-recognized that elution volumes greater than unity can be obtained for both carbohydrate- (33) and polyacrylamide- (17) based permeation supports. It is recommended that eluants containing, for example, 0.1 M NaCl should be used to prevent adsorption phenomena affecting the separation. The use of polyacrylamide-based gels is also recommended for optimum resolution at high temperatures (e.g. 65°C) (19) or when samples of biological origin are used (17) in order to prevent elution of extraneous carbohydrate material due to disruption of the gel matrix. Such a system can be operated using automated assay detection systems for up to 1 year with no loss of resolution (19).

The use of oligosaccharide fractionation by gel permeation is now recommended as a superior method for the characterization of starch hydrolysates (2) particularly in the determination of dextrose equivalent (34) based on the actual composition of oligosaccharides present rather than on a single gross determination (see Section 2).

# 5. HIGH PERFORMANCE LIQUID CHROMATOGRAPHY

H.p.l.c., characterized by small particle sizes (<25 μm), narrow bore columns, high inlet pressures and short analysis times, uses many of the same separation principles described for low pressure column chromatography. Since its inception in the mid 1970's the number of methods and types of packing material have expanded rapidly. Much of the earlier work on oligosaccharide analysis has been reviewed (24) and the theory, separation modes and instrumentation described (35) and the developments are such that a number of the low pressure chromatographic systems are now being replaced, particularly in the food industry for the analysis of the simpler mixtures. Details of the experimental approach can be found in Chapter 1.

## 5.1 Adsorption chromatography

Adsorption, or normal phase, chromatography relies on the surface hydroxyl groups of silica (and to a lesser extent alumina) which can interact with solutes and affect a separation on account of the different strengths of interaction. The separation of neutral oligosaccharides cannot be carried out conveniently by this method although limited separations can be achieved in water (36) or using solvent mixtures (37). The method is, however, well suited to the analysis of derivatives of oligosaccharides of low d.p. using non-aqueous eluants (see *Table 1*) particularly when ultrasensitive (10−100 ng level) detection is required (see Section 5.6).

## 5.2 Reversed-phase chromatography

Reversed-phase chromatography packings are characterized by hydrocarbon chains bonded to the surface of the silica matrix. Whilst chain lengths can range from 1 to 22 carbon atoms the most popular are the 18 carbon atom chain (octadecylsilyl silica) and the eight carbon chain (octylsilyl silica). The essential criterion responsible for separation is the interaction of the packing with polar materials. Using aqueous solutions or solvents of medium polarity the more polar species elute first and as the polarity decreases the more tightly bound less polar species are eluted.

The use of high concentrations of organic solvents in the mixtures of water and solvent used as eluant gives rise to problems of solubility of oligosaccharides. Nevertheless oligosaccharides up to d.p. 30 contained in wood extracts have been fractionated in under 30 min using a gradient of 70−62.5% acetonitrate in water (42) although the separation between the first five members of the series (d.p. 1−5) was very poor. Use of fully acetylated oligosaccharides overcomes problems of solubility and has resulted in the fractionation of malto-oligosaccharides up to d.p. 30 in about 150 min (43) using an exponential gradient of acetonitrile (10−70%) in water.

**Table 1.** Derivatives used for the adsorption chromatography of oligosaccharides.

| Derivative | Eluant | Reference |
|---|---|---|
| Benzoate | Hexane-ethyl acetate (85:15) | 38 |
| 4-Nitrobenzoate | Hexane-acetonitrile-chloroform (65:20:15) | 39 |
| Benzyloxime-perbenzoate | Hexane-dioxane (80:20) | 40 |
| Phenyldimethylsilyl ethers | Hexane-ethyl acetate (197:3) | 41 |

## 5.3 **Bonded-phase chromatography**

By far the most frequently used systems for separation of oligosaccharides are those using chemically bonded phases which fractionate materials on the basis of their relative affinities for the mobile phase and the bonded phase. Supports containing a variety of cyano- or amino-bonded phases have been used in the past but the two most important types of column are those containing the aminopropyl bonded phase (44) and a hybrid phase of cyano- and amino-derivatives, i.e. the Partisil PAC column (45) with the aminopropyl-bonded column being regarded by some workers as the ultimate in column technology for oligosaccharide analysis (46).

Separation of series of oligosaccharides from, for example, hydrolysed starch can readily be achieved with up to d.p. 10 being separated in 15−20 min using acetonitrile-water eluants containing 35−40% water (see *Figure 8*). Increasing the water content to 45% can increase the number of detectable oligosaccharides up to about d.p. 15. However very high molecular weight materials cannot be analysed due to prolonged retention times and solubility problems in the acetonitrile-water eluants and for a full analysis separation by gel permeation or ion exchange is also required.

Prolonged use of bonded amine columns is accompanied by loss of performance through loss of the bonded phase and to overcome this shortcoming and to introduce different degrees of selectivity into the system the use of in-situ modification of the silica has been developed (47). Instead of using an amine phase bonded to the silica support a diamine or polyamine is added to the eluant which results in a dynamic equilibrium between an amine phase coating the silica and that in the eluant. Separations are similar to those obtained using bonded amine columns with separations of up to d.p. 20−25 being achieved in 45 min (see *Figure 9*) using eluants containing 50% water in acetonitrile to which the modifier has been added at the 0.01% (v/v) level. Even at this low modifier concentration it is imperative to use a presaturation column in the system prior to the injector to protect the analytical column packing from dissolution. Whilst diamine modifiers give the optimum resolution between oligosaccharides of different d.p. polyamine modifiers give improved selectivity and separa-

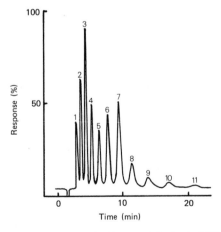

**Figure 8.** Separation of starch-derived oligosaccharides on a Spherisorb S5 NH₂ column (1,2,3, etc., refer to the d.p. of the oligosaccharide).

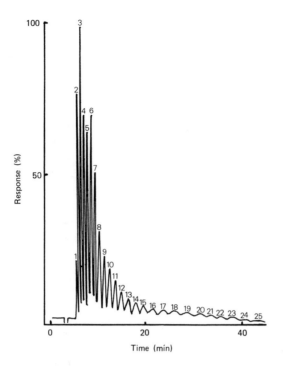

**Figure 9.** Separation of starch-derived oligosaccharides on an *in-situ* modified silica column using 1,4-diaminobutane (0.01% v/v) as modifier. (1,2,3, etc. refer to the d.p. of the oligosaccharide).

tion of oligosaccharides of the same d.p. such that the presence of, for example, isomaltose in starch derived oligosaccharides can be determined (46).

## 5.4 Ion-exchange chromatography

High performance ion-exchange chromatography utilizes the same mechanisms for separation described in Section 4.1. Anion-exchange resins have not been developed to their full extent for oligosaccharide analysis but examples of styrene-divinylbenzene matrices in the sulphate form have been reported (48) for the separation of the simple disaccharides using 80−90% ethanol in water eluants. Cation resins in the lithium form have been used for similar analyses using 90% ethanol in water eluants (49).

The use of 4 and 8% cross-linked cation-exchange resins in the calcium (50,51) or silver (52) form have been used to provide a rapid separation of oligosaccharides of up to d.p. 6−8 for calcium counter ions (51) or d.p. 8−12 for silver counter ions (52) depending on the size of the chromatographic column and the time of analysis. The major drawbacks to the use of such systems include the compressibility of the gel matrix; extended analysis times; efficiency losses of the order of 50% for a doubling of the flow-rate; the need for high temperature (85°C) operation; and the need for specialized regeneration and re-packing of contaminated columns. However these drawbacks are offset by the ability to obtain a total analysis of material applied to the column (see *Figure 10*) and the use of water as the only eluant.

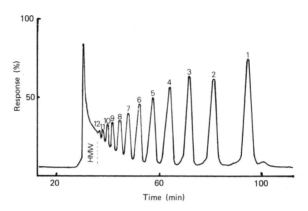

**Figure 10.** Separation of starch-derived oligosaccharides by high performance ion-exchange chromatography using 4% cross-linked cation exchange resin with silver counter ions (1,2,3, etc. refer to the d.p. of the oligosaccharide; HMW, high molecular weight material above d.p. 12).

## 5.5 Gel permeation chromatography

The development of non-compressible matrices for g.p.c. analysis of water soluble materials has not reached the degree of sophistication available for the organic eluant compatible matrices. Consequently there has been no direct replacement for the cross-linked polysaccharide or polyacrylamide materials used for traditional gel permeation analysis of oligosaccharides. Some advances have, however, been made with the development of silica matrices deactivated by the chemical bonding of an organic ether stationary phase to provide a hydrophilic surface. The currently available materials have fractionation ranges which extend down to molecular weights of about 2000 (i.e. d.p. 10−12) whilst unmodified silica with 6 nm pore size can extend the fractionation range down to molecular weights of about 1000 (i.e. d.p. 5−6). Even with the modified materials adsorption effects are present and elution with ionic buffers (0.1 M) is recommended within the pH range 2−7. Oligosaccharides up to d.p. 6 from partially degraded hyaluronic acid have been resolved but a complicated three-column arrangement was required (53).

Water compatible hydroxylated polyether-based matrices have been developed (for example the TSK PW series) which overcome some of the disadvantages of silica-based materials and have fractionation ranges which are comparable to the cross-linked polysaccharide and polyacrylamide gels. Such materials are less rigid than the silica-based materials and therefore require lower operating pressures. Their lower selectivity compared to silica-based matrices is offset by their increased stability towards alkaline pH (up to pH 12). Whilst analysis times are of the order of one third to one tenth that of traditional low pressure gel permeation analysis the separatoins are inferior to those currently obtainable by ion-exchange chromatography (Section 5.4) and consequently little emphasis is placed on high performance gel permeation chromatography for oligosaccharide fractionation.

## 5.6 Detection

Detection of oligosaccharides eluting from h.p.l.c. columns is the biggest challenge and weakest link in the analysis of oligosaccharides. Non-specific detectors such as the refractive index detector are used routinely, but their lack of sensitivity and restriction to isocratic elution (i.e. single non-gradient eluants) severely limits the development of improved techniques. Low wavelength u.v. detection (below 210 nm) has been shown to have comparable sensitivity whilst allowing the use of limited gradient elution (54) but the requirement for ultra-pure solvents renders the method impractical for routine use. A mass detector has been developed which allows gradient elution to be performed and has been used for monosaccharide analysis (55) but the use of high water contents in the eluants needed for oligosaccharide analysis results in instability of the evaporation system which is an integral part of the detection system.

When sensitivity in the sub-microgram range is required the only method readily available to date is the use of pre-column derivitization (see Section 5.1) with separation via adsorption chromatography and detection via u.v. monitoring. This method has been extended to the use of bonded amine columns using u.v.-absorbing substituted oximes such as the *O*-methyl, *O*-benzyl and *O*-(4-nitrobenzyl) derivatives (56) or fluorescent derivatives using 2-aminopyridine or 7-amino-1-naphthol to form Schiff-base complexes (57).

Whilst the use of post-column reaction systems have been developed for monosaccharide analysis (see Chapter 1) such systems have a limited applicability to oligosaccharide analysis due to the reliance on reducing sugar assays. Attempts to use strong cation-exchange resins (in the protonated form) to hydrolyse the glycosidic bonds in oligosaccharides to give complete conversion to monosaccharides after chromatographic separation has been reported (58) with the resulting monosaccharides being detected as reducing compounds.

The problems of removing all traces of eluant after chromatographic separation are being resolved and are allowing the mass spectrometer to be developed as an ultra-sensitive h.p.l.c. detector (59) which could have many applications in oligosaccharide analysis.

## 6. GAS CHROMATOGRAPHY

Despite the expansion of h.p.l.c. techniques in the past five years, gas chromatography (g.c.) still continues to have a place in oligosaccharide analysis for both structural studies to determine the component monosaccharide residues and position of inter-residue glycosidic bonds (see Chapter 3) and the routine analysis and quantitation of oligosaccharides. Whilst the determination of oligosaccharides by g.c. is normally restricted to the analysis of disaccharides as an extention of monosaccharide analysis (see Chapter 1) methods have been reported for the fractionation of oligosaccharides up to d.p. 6−7 (60). The choice between h.p.l.c. and g.c. is not clear cut with g.c. being at least 10 times more sensitive and producing shorter separation times than h.p.l.c. techniques. To offset these advantages the additional time required to prepare the sample and produce the required derivative (see Chapter 1) with no partially derivatized contaminants makes single analyses very unattractive. The final decision as to whether to use h.p.l.c. or g.c. frequently depends not on the relative merits of the method but on the nature of a particular sample (61). Typical samples which are better suited to g.c. analysis are those containing trace amounts of oligosaccharides as in plant tissues or where the low d.p. oligosaccharides are present in small amounts relative to high contents of disaccharides and monosaccharides such as found in high DE glucose syrups (60) or adulterated honey (62).

## 7. MASS SPECTROMETRY

While mass spectrometry has traditionally been of major use when coupled to g.c. columns to provide structural information on oligosaccharides and polysaccharides (see Chapter 3) the development of several new ionization techniques now allows the direct analysis of underivatized oligosaccharides. Field desorption, in which ions are produced by thin wire emitters on which the sample is deposited, is a very soft ionization technique capable of producing molecular ions or simplified spectra (63). The most recent technique, fast-atom bombardment, produces ions via bombardment of the sample with high-energy rare-gas atoms. Structural information can be obtained in addition to molecular weight for underivatized oligosaccharides of d.p. up to 20 and beyond.

## 8. NUCLEAR MAGNETIC RESONANCE SPECTROSCOPY

Nuclear magnetic resonance (n.m.r.) spectroscopy has been developed into a very useful non-destructive technique for the determination of oligosaccharide structures. $^{13}$C-N.m.r. not only gives information on the anomeric configuration of the carbohydrate residues but also provides information on the composition of component monosaccharides, their sequence and the overall conformation of the molecule (see Chapter 3). $^{1}$H-N.m.r., whilst being a more sensitive technique, does not provide as much structural information due to incomplete separation of the proton resonance signals obtained for carbohydrate molecules. It can, however, be used to quantitate individual oligosaccharide contents in situations where the unique selectivity of the method overcomes interference problems encountered when using other spectroscopic methods of great sensitivity but inferior selectivity. Determination of substituent groups on carbohydrate residues can also be made without the problems encountered with the traditional degradation techniques (64). It is outside the scope of this chapter to describe the various methods available for recording spectra and the methods by which spectra can be interpreted. These subjects have been reviewed in depth recently with particularly emphasis being given to oligosaccharide analysis (65,66).

With the advent of Fourier-transform techniques development is underway to prepare on-line h.p.l.c. detectors (67) which will provide valuable structural information on components in mixtures rather than the more normal determination of net properties of a component. This non-destructive detector will allow its use in conjunction with other detectors or fraction collectors if preparative scale chromatography is used.

## 9. INFRA-RED SPECTROSCOPY

Infra-red spectra of carbohydrates are complex and the method is usually restricted to the identification of specific structural features, particularly when monitoring of chemical reactions is required with interpretation of spectra normally by comparison with published spectra or spectra obtained for a standard reference compound. The advent of infra-red reflectance spectroscopy has led to the development of analysers capable of detecting total carbohydrate contents in aqueous solutions and complex mixtures and formulations (68,69). By choice of selected resonance frequencies it is possible to introduce a degree of selectivity into the determination such that the concentration of D-glucose in starch suspensions or specific oligosaccharides such as lactose can be measured.

## 10. REFERENCES

1. Dziedzic,S.Z. and Kearsley,M.W. (eds.) (1984) *Glucose Syrups: Science and Technology*, Elsevier Applied Science, London.
2. Kennedy,J.F., Noy,R.J., Stead,J.A. and White,C.A. (1985) *Starch/Stärke*, **37**, 343.
3. Dische,Z., Shettles,L.B. and Osnos,M. (1949) *Arch. Biochem.*, **22**, 169.
4. Dubois,M., Gilles,K.A., Hamilton,J.K., Rebers,P.A. and Smith,F. (1956) *Analyt. Chem.*, **28**, 350.
5. Svennerholm,L. (1956) *J.Neurochem.*, **1**, 42.
6. Roe,J.H. (1955) *J. Biol. Chem.*, **212**, 335.
7. Lane,J.H. and Eynon,L. (1923) *J. Chem. Soc. Ind.*, **42**, 32T.
8. Egan,H., Kirk,R.S. and Sawyer,R. (1981) *Pearson's Chemical Analysis of Foods*, 8th edition, Churchill Livingstone, London, p. 150.
9. Bernfeld,P. (1955) in *Methods in Enzymology*, Colowick,S.P. and Kaplan,N.D. (eds), Academic Press, London and New York, Vol. 1, p. 149.
10. Lindner,W.A., Dennison,C. and Quicke,G.V. (1983) *Biotechnol. Bioeng.* **25**, 377.
11. Robyt,J.F. and Whelan,W.J. (1972) *Analyt. Biochem.*, **45**, 510.
12. Somogyi,M. (1952) *J. Biol. Chem.*, **195**, 19.
13. Breuil,C. and Saddler,J.N. (1985) *Enzyme Microb. Technol.*, **7**, 327.
14. Kidby,D.K. and Davidson,D.J. (1973) *Analyt. Chem.*, **55**, 321.
15. Osborne,D.R. and Voogt,P. (1978) *The Analysis of Nutrients in Food*, Academic Press, London.
16. Mullings,R. and Parish,J.H. (1984) *Enzyme Microb. Technol.*, **6**, 491.
17. White,C.A. and Kennedy,J.F. (1979) *Clin. Chim. Acta*, **95**, 369.
18. Kennedy,J.F. and Fox,J.E. (1980) *Methods Carbohydrate Chem.*, **8**, 3.
19. Kennedy,J.F. and Fox,J.E. (1980) *Methods Carbohydrate Chem.*, **8**, 13.
20. Buffa,M., Congiu,G., Lombard,A. and Tourn,M.L. (1980) *J. Chromatogr.*, **200**, 309.
21. Kennedy,J.F., White,C.A. and Riddiford,C.L. (1979) *Starch/Stärke*, **31**, 235.
22. Damonte,A., Lombard,A., Tourn,M.L. and Cassone,M.C. (1971) *J. Chromatogr.*, **60**, 203.
23. Stahl,E. and Kaltenbach,U. (1961) *J. Chromatogr.*, **5**, 351.
24. Heyraud,A. and Rinaudo,M. (1981) *J. Liquid Chromatogr.*, **4**, 175.
25. French,D., Robyt,J.F., Weintraub,M. and Knock,P. (1966) *J. Chromatogr.*, **24**, 68.
28. Havlicek,J. and Samuelson,Ö. (1972) *Carbohydr. Res.*, **22**, 307.
27. Barker,S.A., Hatt,B.W., Kennedy,J.F. and Somers,P.J. (1969) *Carbohydr. Res.*, **9**, 327.
28. Havlicek,J. and Samuelson,Ö. (1972) *Carbohydr. Res.*, **22**, 307.
29. Jonsson,P. and Samuelson,Ö. (1967) *J. Chromatogr.*, **26**, 194.
30. Kennedy,J.F., Cabral,J.M.S. and Kalogerakis,B. (1985) *Enzyme Microb. Technol.*, **7**, 22.
31. White,C.A., Kennedy,J.F., Lombard,A. and Rossetti,V. (1985) *Br. Polymer J.*, **17**, 327.
32. Kennedy,J.F., Stevenson,D.L. and White,C.A. (1985) *Cellulose Chem. Technol.*, **19**, 505.
33. Kennedy,J.F. (1972) *J. Chromatogr.*, **69**, 325.
34. Kiser,D.L. and Hagy,R.L. (1979) In *Liquid Chromatographic Analysis of Food and Beverages*, Charalambous,G. (ed.), Academic Press, New York, Vol. 2, p. 263.
35. Macrae,R. (ed.) (1982) *HPLC in Food Analysis*, Academic Press, London.
36. Rocca,J.F. and Rouchousse,J. (1976) *J. Chromatogr.*, **117**, 216.
37. Iwata,S., Narui,T., Takahashi,K. and Shibata,S. (1984) *Carbohydr. Res.*, **133**, 157.
38. White,C.A., Kennedy,J.F. and Golding,B.T. (1979) *Carbohydr. Res.*, **76**, 1.
39. Nachtmann,F. and Budna,K.W. (1977) *J. Chromatogr.*, **136**, 279.
40. Thompson,R.M. (1978) *J. Chromatogr.*, **166**, 201.
41. White,C.A., Vass,S.W., Kennedy,J.F. and Large,D.G. (1983) *J. Chromatogr.*, **264**, 99.
42. Noel,D., Hanai,T. and D'Amboise,M. (1979) *J. Liquid Chromatogr.*, **2**, 1325.
43. Wells,G.B. and Lester,R.L. (1979) *Analyt. Biochem.*, **97**, 184.
44. Schwarzenbach,R. (1976) *J. Chromatogr.*, **117**, 206.
45. Rabel,F.M., Caputo,A.G. and Butts,E.T. (1976) *J. Chromatogr.*, **126**, 731.
46. Folkes,D.J. and Taylor,P.W. (1982) In *HPLC in Food Analysis*, Macrae,R. (ed.), Academic Press, London, p. 149.

47. White,C.A., Corran,P.H. and Kennedy,J.F. (1980) *Carbohydr. Res.*, **87**, 165.
48. Oshima,R., Takai,N. and Kumanotani,J. (1980) *J. Chromatogr.*, **192**, 452.
49. Boykins,R.A. and Liu,T.-Y. (1981) In *Biological and Biomedical Applications of Liquid Chromatography*, Haw,G.L. (ed.), Marcel Dekker, New York, Vol. 3, p. 347.
50. Ladisch,M.R. and Tsao,G.T. (1978) *J. Chromatogr.*, **166**, 85.
51. Fitt,L.E., Hassler,S. and Just,D.E. (1980) *J. Chromatogr.*, **187**, 381.
52. Scobell,H.D. and Brobst,K.M. (1981) *J. Chromatogr.*, **212**, 51.
53. Kundsen,P.J., Eriksen,P.B., Fenger,M. and Florentz,K. (1980) *J. Chromatogr.*, **187**, 373.
54. Binder,H. (1980) *J. Chromatogr.*, **189**, 414.
55. Macrae,R. and Dick,J. (1981) *J. Chromatogr.*, **210**, 138.
56. Chen,C.-C. and McGinnis,G.C., (1981) *J. Chromatogr.*, **122**, 322.
57. Coles,E., Reinhold,V.N. and Carr,S.A. (1985) *J. Chromatogr.*, **139**, 1.
58. Vrátný,P. and Ouhrabková,J. (1980) *J. Chromatogr.*, **191**, 313.
59. Chapman,J.R., (1985) *Inst. Food Sci. Technol. Proc.*, **18**, 59.
60. Folkes,D.J. and Brookes,A. (1984) In *Glucose Syrups: Science and Technology*, Dziedzic,S.Z. and Kearlsey,M.W. (eds), Elsevier Applied Science, London, p. 197.
61. Brobst,K.M. and Scobell,H.D. (1982) *Starch/Stärke*, **34**, 117.
62. Doner,L.W., White,J.W. and Phillips,J.G. (1979) *J. Assoc. Off. Anal. Chem.*, **62**, 186.
63. Schutten,H.-R. and Lehmann,W.D. (1980) *Trends Biochem. Sci.*, **5**, 142.
64. Kennedy,J.F., Stevenson,D.L., White,C.A., Tolley,M.S. and Bradshaw,I.J. (1984) *Br. Polymer J.*, **16**, 5.
65. Coxon,B. (1980) In *Developments in Food Carbohydrates - 2*, Lee,C.K. (ed.), Applied Science, London, p. 351.
66. Rathbone,E.B. (1985) In *Analysis of Food Carbohydrate*, Birch,G.G. (ed.), Elsevier Applied Science, London, p. 149.
67. Dorn,H.C. (1984) *Analyt. Chem.*, **56**, 747A.
68. Kennedy,J.F., White,C.A. and Browne,A.J. (1985) *Food Chem.*, **16**, 115.
69. Kennedy,J.F., White,C.A. and Browne,A.J. (1985) *Food Chem.*, **18**, 95.

CHAPTER 3

# Neutral polysaccharides

JOHN H.PAZUR

## 1. INTRODUCTION

Polysaccharides are polymers of monosaccharide residues joined together by glycosidic bonds which are formed by the elimination of the elements of water between the hemiacetal hydroxyl group of one residue and a primary or secondary hydroxyl group of an adjacent residue (1). Polysaccharides may be of a linear, branched or occasionally cyclic structure, may contain a small or a large number of residues and may be homogeneous or heterogeneous in monosaccharide types. Originally polysaccharides were named to reflect biological sources, properties or metabolic functions. At the present time, polysaccharides are named on the basis of the monosaccharide units comprising the polymers (2). In accord with this system, a polysaccharide consisting of a single type of monosaccharide is named by replacing the 'ose' ending of the constituent monosaccharide unit with the suffix 'an' as for example, glucan for a polysaccharide of glucose. In the case of a polysaccharide with more than one type of residue the name is determined from the nature of the structural unit of the main chain and the prefix names of the other units, as for example, fuco-manno-galactan. In the case of a polysaccharide with no main chain, the compound is named by listing the prefix names of all of the constituent units alphabetically followed by the general term glycan, as for example galacto-gluco-manno-glycan.

The terms glycan and polysaccharide are equivalent terms. Polysaccharides which are composed of one type of monosaccharide unit are called homopolysaccharides or homoglycans, while those composed of two or more types of units are called hetero-polysaccharides or heteroglycans. The determination of the structure of the two types of polymers may necessitate the use of different strategies and analytical procedures. The complexity of the monosaccharide sequence of heteropolysaccharides, while not as great as for the amino acid sequence of proteins, nevertheless can be involved because heteropolysaccharides may possess several types of glycosidic bonds as well as several types of monosaccharide units. If the formation of glycosidic bonds of polysaccharides occurred in a random fashion in nature, the structure of polysaccharides could be even more complex. However, a completely random formation of glycosidic bonds does not occur and therefore most polysaccharides possess repeating units of relatively uniform structure. The pathways of synthesis of polysaccharides are under genetic regulation and the types of the synthesizing enzymes, the reaction mechanisms and the rates of synthesis and degradation are therefore predetermined.

Naturally occurring polysaccharides vary greatly in molecular weight ranging from 5000 to several million daltons. The molecular structure and architecture are also diverse.

These structures can be divided into three basic types, a linear structure in which the units are joined by a uniform type of glycosidic linkage to form long linear chains, a substituted linear structure in which short side chains of single or oligosaccharide units are attached to the main chain and a branched structure in which long side chains of monosaccharide residues are attached randomly to the main chain and to side chains. Polysaccharides with structures intermediate to the above basic types have been isolated and characterized.

The structural features of a polysaccharide to be determined for the elucidation of the complete molecular structure are as follows: the types of monosaccharide residues, the D or L configuration of the monosaccharide residues, the number of monosaccharide residues in a polysaccharide, the positions of glycosidic linkages between monosaccharide residues, the ring structure of the monosaccharide residues, the sequence of monosaccharide residues, and the $\alpha$ or $\beta$ configuration of the glycosidic bonds.

Theoretical aspects of the analytical techniques used in the structural analysis of polysaccharides, descriptions of experimental protocols for important techniques and examples of the type of results which may be obtained are outlined in this chapter. The analytical methods commonly used are methylation analysis, periodate oxidation, partial acid hydrolysis, acetolysis, alkaline degradation, enzymic assays and immunological reactivities. The methods utilize paper chromatography, gas liquid chromatography (g.l.c.), mass spectrometry (m.s.), polarimetry, circular dichroism (c.d.) absorption, nuclear magnetic resonance (n.m.r.) spectrometry, affinity chromatography and antibody-antigen complex formation.

## 2. ISOLATION METHODS

Polysaccharides are present in the seeds, stems and leaves of plants in various tissues and fluids of animals, in shells of insects and crustaceans, and in cell walls and extra cellular fluids of bacteria, yeasts and fungi. The two major functions of polysaccharides in these biological materials are first as structural elements for rigidity of the tissues and second as energy sources for developing embryos and for metabolic reactions of living entities. The polysaccharides do not occur in pure form but rather as heterogeneous mixtures with other cellular components. Accordingly the polysaccharides must first be isolated and purified to homogeneity by suitable methods. Adequate criteria of purity must also be obtained before structural analysis can be undertaken. Many structural studies have been performed on preparations of questionable purity and the value of such information is minimal. With regard to purification and criteria of purity, there is no single purification method that can be used and there is no single measure of purity that can be applied. It is generally agreed that a polysaccharide can be considered to be pure if it can be isolated by two different procedures and the resulting preparations possess the same chemical and physical properties.

Of the many stereoisomers of monosaccharides which are possible, only a few occur as constituents of polysaccharides and these are listed in *Table 1*. Starch, a homoglycan of glucose with $\alpha$-glycosidic linkages, and cellulose, also a homoglycan of glucose with $\beta$-glycosidic linkages are the most abundant and most widely distributed. Starch is a heterogeneous mixture of two polymers, amylose and amylopectin. The amylose is a glucose polymer in which the glucose units are joined by $\alpha$-D-$(1 \rightarrow 4)$ linkages to form

**Table 1.** Common monosaccharide constituents of polysaccharides.

| Type | Compound |
|------|----------|
| Pentoses | D-xylose, L-arabinose |
| Hexoses | D-glucose, D-mannose, D-galactose, L-galactose, D-fructose |
| Hexosamines | N-acetyl-D-glucosamine, N-acetyl-D-galactosamine |
| Uronic acids | D-glucuronic acid, D-galacturonic acid, D-mannuronic acid |
| Deoxy hexoses | L-rhamnose, L-fucose, 6-deoxy-L-talose |

a linear molecule, and the amylopectin is also a glucose polymer in which the units are joined by $\alpha$-D-$(1\rightarrow4)$ and $\alpha$-D-$(1\rightarrow6)$ linkages. Since some of the glucose units of amylopectin are substituted at the 4 and 6 positions a branched structure results for this component. Starch is the major polysaccharide in cereal grains and functions as an energy source for the developing embyro and as an energy source for humans and animals which consume the starch. Starch also has many industrial uses in the manufacture of paper, textiles, adhesives, films, alcohol, organic acids and other products. Cellulose occurs in the woody tissues of all plants and is nutritionally important in ruminant animals but not in humans. Cellulose is used industrially in the manufacture of paper, textiles and related products. Two other classes of polysaccharides of plant origin possess economic value and these are pectic substances consisting of a polysaccharide complex of arabinose, galactose and galacturonic acid and the plant gums consisting of various combinations of arabinose, rhamnose, galactose, glucuronic acid and other monosaccharides. Chitin, a polysaccharide from insects, crustaceans and fungi occurs in large amounts in nature and is composed of *N*-acetyl glucosamine residues joined by $\beta$-D-$(1\rightarrow4)$ linkages. A major industrial use for this polysaccharide has not yet been found. Numerous types of microbial polysaccharides with unique structures have been isolated and industrial uses have been found for some of these.

Because of the wide variations in chemical and physical properties of polysaccharides the isolation procedures vary greatly. For the isolation of starch, the grain is steeped to remove the seed coat and germ and the resulting material is ground, suspended in water and subjected to differential centrifugation. Cellulose is highly insoluble and procedures for removing other substances associated with the cellulose have been devised. One method utilizes extraction of the plant residues with alkaline sodium chlorite solutions in which the non-cellulosic components are dissolved and are removed by filtration. The cellulose component is present in the residue but is still contaminated with other polymers. Additional treatment of this residue is needed in order to obtain cellulose sufficiently pure for structural analysis. The hemicelluloses can be removed to a large extent by extraction with relatively concentrated sodium hydroxide. The undissolved substance is essentially pure cellulose. The preparation of polysaccharides from insects, animal tissues and microorganisms will require special isolation techniques. A listing of the procedures based on different principles for isolating polysaccharides is presented in *Table 2*.

A number of criteria of purity must be used to establish the purity of a polysaccharide. Procedures which can be used are:

(i)     constancy in monosaccharide composition on repeated purification;
(ii)    quantitative determination of a unique structural constituent following each step

**Table 2.** Methods for the isolation of polysaccharides.

| | | |
|---|---|---|
| 1. | Solubility | |
| | (a) | Fractional precipitation |
| | (b) | Fractional solution |
| | (c) | Distribution between immiscible solvents |
| 2. | Ultracentrifugation | |
| | (a) | Model E ultracentrifuge |
| | (b) | Density gradient techniques |
| 3. | Membrane ultrafiltration | |
| 4. | Chromatography | |
| | (a) | Adsorption |
| | (b) | Ion exchange |
| | (c) | Partition chromatography |
| 5. | Gel filtration | |

of the purification process;
(iii)  the ratio of monosaccharide constituents on repeated purification;
(iv)  sedimentation rate on ultracentrifugation;
(v)  behaviour on gel filtration and ion exchange chromatography.

For the first three methods, suitable analytical techniques must be devised. For sedimentation methods, ultracentrifuges must be available in which sedimentation rates of the polysaccharides can be determined. Polysaccharides which sediment as a single symmetrical peak are most likely homogeneous in molecular size. Multi-peaks, of course, indicate molecular heterogeneity. A preparation may consist of two polysaccharides of identical molecular weights and a single peak would not indicate structural homogeneity.

The behaviour of the polysaccharides on gel filtration or in ion-exchange chromatography yields information about the purity of a polysaccharide preparation. In these procedures the preparation is passed through a column of the gel or adsorbent and the eluates from the column are analysed for carbohydrate. A symmetrical elution peak is indicative of homogeneity. The technique of gel filtration has become a popular method for assessing homogeneity and determining molecular weights. By using polysaccharides of known molecular size as markers, molecular weights of unknowns can be determined. Ultrafiltration through calibrated membranes is another method being used to assess homogeneity in polysaccharide preparations.

## 3. IDENTIFICATION OF MONOSACCHARIDES

In the structural analysis of polysaccharides it is first necessary to determine the types of the monosaccharide residues which constitute the polysaccharide. With the advent of partition chromatography, the determination of the monosaccharide constituents has been considerably simplified. For chromatographic analysis of the monosaccharides, the polysaccharide must first be hydrolysed to constituent units. A satisfactory method for hydrolysis is as follows.

(i)   Dissolve a sample of 2−5 mg of the polysaccharide in 0.1−0.25 ml of 2 M HCl and heat at 100°C for periods of 2−5 h in a sealed tube.

(ii)  Analyse the hydrolysate for monosaccharide components by a partition method using paper, thin layer, automated ion-exchange, g.l.c. or high performance liquid chromatography (h.p.l.c.).

(iii) Analyse suitable standards of monosaccharides or derivatives by the same chromatographic method.

The selection of a chromatographic method will depend on the availability of equipment, and standards and on the supply of polysaccharide. Two very useful procedures are paper chromatography (3,4, see Chapter 1, Section 4) and automated ion-exchange chromatography (5).

A photograph of a paper chromatogram showing the monosaccharides liberated during acid hydrolysis of gum arabic in 2 M HCl is reproduced in *Figure 1*. It is noted in the

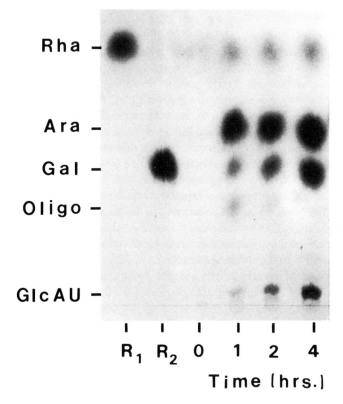

**Figure 1.** A paper chromatogram showing reference compounds and the products of acid hydrolysis of gum arabic as a function of time: $R_1$, rhamnose; $R_2$, galactose; oligo, oligosaccharide and others are standard abbreviations for monosaccharides. The solvent used is *n*-butyl alcohol/pyridine/water 6:4:3 v/v/v, see Chapter 1, Section 4.

figure that rhamnose, arabinose, galactose and glucuronic acid are the monosaccharide constituents of this polysaccharide. The concentrations of monosaccharides can be estimated from the intensity of the spots on the chromatogram. By utilizing standards of several concentrations and comparing colour intensities, quantitative values for the monosaccharides in the hydrolysate can be obtained. It should be noted in *Figure 1* that the arabinose spot is very intense and is present at the early stages of hydrolysis. Such behaviour is indicative of a highly labile linkage and a furanose ring structure for arabinose units. An oligosaccharide fragment also appears in the early stages of hydrolysis and disappears on prolonged hydrolysis but the nature of this oligosaccharide was not investigated. The unequivocal indentification of the monosaccharides will involve the recovery of the monosaccharides from the hydrolysate in sufficient quantities for measurement of physical constants and the preparation and characterization of derivatives.

Other methods of monosaccharide analysis are described in Chapter 1 and reference 6.


## 4. DETERMINATION OF D AND L CONFIGURATION OF COMPONENT MONO-SACCHARIDES

The D or L configuration of the monosaccharide residues of a polysaccharide is determined by appropriate measurements on the monosaccharides liberated by acid or enzyme hydrolysis of the polysaccharide. The monosaccharides are isolated in pure form by a chromatographic method. The specific rotations, c.d. absorptions or enzymic susceptibilities of the monosaccharides are measured. By comparison of the specific rotation of the isolated compounds with that of known monosaccharides or with literature values the D or L configuration of the unknown compound can be established. Comparisons of c.d. absorptions of the unknown with that for known compounds are made and the configuration is established. The susceptibility of the monosaccharides to oxidation, isomerization or phosphorylation by enzymes can be utilized to establish configuration type. The latter method is dependent on the availability of enzymes with the desired specificity for the D or L isomer.

The methods for the determination of the configurations of monosaccharide units of a polysaccharide are illustrated for a polysaccharide composed of 6-deoxy-talose, rhamnose, galactose and glucuronic acid from *Streptococcus bovis*. The molecular structure of this polysaccharide has been determined by use of a combination of analytical methods and the structure of the repeating unit is shown in Formula 1 (7).

$$\rightarrow 3)\text{-}6\text{-deoxy-L-Tal}p\text{-}(1\rightarrow 3)\text{-D-Gal}p\text{-}(1\rightarrow 3)\text{-L-Rha}p\text{-}(1\rightarrow 2)\text{-L-Rha}p\text{-}(1\rightarrow$$

$$4$$
$$\uparrow$$
$$1$$

$$\text{D-GlcA}p$$

**Formula 1**

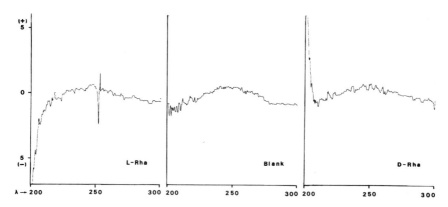

**Figure 2.** Circular dichroism absorption curves for stereoisomers of rhamnose.

(i)     Hydrolyse 100 mg of the polysaccharide in 2 ml of 0.1 M HCl in a boiling water bath for 3 h.

(ii)    Analyse the hydrolysate by qualitative paper chromatography and identify the monosaccharides.

(iii)   Isolate the monosaccharides by preparative paper chromatography as described in Chapter 1, Section 4. From 100 mg of the polysaccharide 12 mg of 6-deoxy-talose, 10 mg of rhamnose, 15 mg of galactose and 8 mg of glucuronic acid were obtained.

(iv)    Use appropriate amounts of the isolated monosaccharide to prepare 1.5 ml solution containing 0.01%, 0.009%, 0.005% and 0.01% of 6-deoxy-talose, rhamnose, galactose and glucuronic acid, respectively.

(v)     Measure the optical rotation of each solution in a 2 dcm polarimeter tube with 1.4 ml capacity in a Rudolph polarimeter. The observed rotation readings were in the range of 0.12° to 0.6°.

(iv)    Calculate the specific rotations from these data: 6-deoxy-talose $[\alpha]_D$ $-17°$ (c 0.01, water) literature value for 6-deoxy-L-talose $-18.9°$; rhamnose $[\alpha]_D =$ $+8°$ (c 0.009, water) literature value for L-rhamnose, $+8.2°$, galactose $[\alpha]_D$ $= +77$ (c 0.005, water) literature value for D-galactose $+80.2$, and glucuronic acid $[\alpha]_D = +35.3$ (c 0.01, water) literature value for D-glucuronic acid $+36.3°$.

The D or L configuration of rhamnose and galactose were also determined by c.d.

(i)     Dissolve 0.2 mg of rhamnose or galactose from the hydrolysate of the polysaccharide in 2 ml of water.

(ii)    Measure the u.v. absorption of 1 ml of these samples in the range of 200–300 nm in a recording spectropolarimeter.

(iii)   Measure the u.v. absorption of standard solutions of D and L rhamnose and D and L galactose and for a water blank. The curves obtained for the D and L rhamnose and for a water blank are reproduced in *Figure 2*. It will be noted that absorptions of L-rhamnose and D-rhamnose yield curves of opposite signs at wave lengths of 200–300 nm. The absorption behaviour of rhamnose isolated from the hydrolysate of the polysaccharide was identical to that for the standard L-rhamnose and the absorption behaviour of the galactose from the polysaccharide was the same as that for standard D-galactose. These data verify the L configuration for the rhamnose and the D configuration for galactose from the polysaccharide.

The configuration of the glucuronic acid from the glycan was also determined by an enzyme method following the conversion of the glucuronic acid to glucose.

(i) Reduce the carboxyl group of the glucuronic acid to a primary alcohol group by reacting 20 mg of polysaccharide with water soluble carbodiimide (20 mg) in 2 ml of water for 6 h and reducing the product with 10 mg of sodium borohydride for 18 h.

(ii) Acidify with acetic acid and remove the borates by evaporation from methanol.

(iii) Dialyse for 24 h and lyophilize to dryness.

(iv) Hydrolyse 5 mg of the product in 0.25 ml of 0.1 M HCl for 2 h in a boiling water bath.

(v) Analyse the hydrolysate by paper chromatography in a solvent system of *n*-butyl alcohol pyridine-water (6:4:3 by vol.) as described in Chapter 1, Section 4.

(vi) Spray the dried chromatogram lightly with a glucose oxidase-peroxidase solution (0.1% glucose oxidase and 0.01% peroxidase in 0.1 M acetate buffer of pH 5.2) and then with alcoholic *O*-tolidine solution (0.2%).

Reference D-glucose and a product in the hydrolysate of the reduced polysaccharide

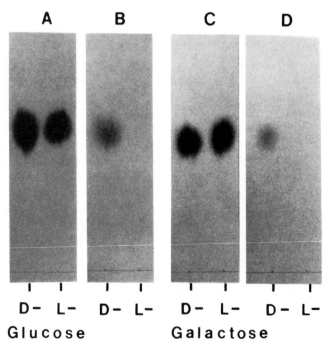

**Figure 3.** Paper chromatographic strips stained for reducing sugars and stained with glucose oxidase or galactose oxidase; **A** and **C** are stained for glucose and galactose by silver nitrate method, **B** is stained with glucose oxidase, peroxidase and *O*-tolidine, and **D** is stained with galactose oxidase, peroxidase and *O*-tolidine.

with identical Rf value to glucose yielded intense blue spots on the chromatogram (8). Reference L-glucose did not yield a positive test with the glucose oxidase. These results and results of galactose oxidase tests are shown in *Figure 3*. Thus, the glucose obtained by reduction of the glucuronic acid is shown to be of the D configuration and therefore the glucuronic acid is also of the D configuration. In the determination of the configuration of galactose by the enzymic method, galactose oxidase replaces the glucose oxidase in the spray solution. This oxidase and the peroxidase are dissolved in 0.1 M phosphate buffer (pH 7). Spray the chromatogram containing D and L galactose and the hydrolysate of the polysaccharide with the galactose oxidase peroxidase spray followed by the *O*-tolidine spray. Stain a duplicate chromatogram to reveal the position of the reducing sugars. It is noted in *Figure 3* that D-glucose but not L-glucose yielded a positive test with glucose oxidase and D-galactose but not L-galactose yielded a positive test with galactose oxidase. Enzymes with other types of stereospecificity can be used to establish the configuration of the monosaccharide units, as for example L-rhamnose isomerase, D-glucokinase, D-galactokinase and D-glucose isomerase.

## 5. DETERMINATION OF THE DEGREE OF POLYMERIZATION

In the elucidation of the structure of polysaccharides it is necessary to calculate the number of monosaccharide residues in a molecule of the polysaccharide. Since the molecular weight of a polysaccharide is not rigidly controlled during biosynthesis, a distribution of molecular weights will exist. Molecular weights are then average molecular weights. Several types of methods have been employed for determining molecular weights, but no single method is ideal. Older methods based on osmotic pressure, viscosity, light scattering and streaming biofringance measurements are not used appreciably at the present time. The newer methods are based on gel fitration, ultracentrifugation and ultrafiltration (9,10).

Ultracentrifugation can be used if the molecular weight of the polysaccharide is reasonably large. Methods based on the use of an analytical ultracentrifuge or preparative ultracentrifuge and density gradient columns are available. In the methods utilizing the analytical ultracentrifuge standard cells and buffers are used and centrifugations are performed at 59 780 r.p.m. The sedimentation of the polymer is observed by use of an elaborate optical arrangement. Other properties of the polysaccharide specifically diffusion coefficient, partial specific volume and the density of the solvent must also be measured. The sedimentation rate can then be calculated and the molecular weight determined (10). In the density gradient centrifugation method (11) gradients are prepared from 5% to 40% solution of sucrose or glycerol and from caesium chloride solutions containing $1.02 - 1.10$ g/ml. The sucrose column can be used only with polysaccharides that can be assayed in the presence of sucrose.

(i)  Place a sample of about 2 mg on top of the gradient column.
(ii)  Place the tubes in a swinging bucket rotor and centrifuge at 65 000 r.p.m. for $10-16$ h.
(iii)  Fractionate the column in 0.1 ml fractions by means of a mechanical fractionator (11).
(iv)  Analyse the eluates for the polysaccharide component by a suitable colorimetric method (6).

(v)     Subject the polysacharides of known molecular weights to identical centrifugation conditions.

(vi)    Measure the distance that the polysaccharides sediment from the top meniscus to the middle of the polysaccharide peak.

(vii)   From an empirical relationship $(D_1/D_2) = (M_1/M_2)^{2/3}$ calculate the molecular weight of the polysaccharide (12). $D_1$ and $M_1$ are distance and molecular weight for the unknown and $D_2$ and $M_2$ for the standard.

In the density gradient centrifugation procedure, the polysaccharide can be easily recovered following centrifugation by combining the fractions containing the polymer, dialysing to remove gradient material and taking to dryness by lyophilization.

Ultracentrifugation of polysaccharides in an analytical centrifuge yields data from which the sedimentation constant and molecular weight can be calculated but recovery of the polysaccharide is more difficult to achieve.

The molecular weights of polysaccharides of low degree of polymerization can be determined by methods based on the quantitative determination of a functional group of the polysaccharide and appropriate standard curves prepared for polysaccharides of known molecular weights. The reducing group of polysaccharides can be measured by colorimetric procedures and can be reacted with reagents containing $^3H$ or $^{14}C$ (13). The colour intensity and radioactivity values of the reaction products with the polysaccharide can be correlated with molecular weights. The carbohydrate residues which constitute the non-reducing end of a polysaccharide can be quantitatively determined and the values used to calculate molecular weights and chain lengths. In one such method, methylate the polysaccharide completely, then hydrolyse to the constituent monosaccharides. The residue at the non-reducing end will yield a unique methylated derivative. For linear polysaccharides the degree of polymerization of linear molecules is calculated from the yield of this derivative. For branched molecules the average chain lengths to the branch points can be determined. In another method for analysis for terminal residues, periodate oxidation is used to obtain data for calculation of molecular weights. In this reaction, formic acid is produced from the residue at the non-reducing end of the polysaccharide and the amount of acid is measured quantitatively by a titration method and used in the calculations.

Recently, gel filtration methods based on the use of gels of various pore sizes have become available and are being used for the determination of molecular weight of polysaccharides (9). In the filtration methods, standard polysaccharides of known molecular weights are needed for calibration curves. Also it is necessary to establish in these procedures that polysaccharides of unknown molecular weight behave in an identical manner as the standards in the gel permeation. Also ultrafiltration methods on filters which retain molecules of specified sizes have become available. Such methods may be advantageous to use for the determination of molecular size of some polysaccharides.

## 6. DETERMINATION OF THE POSITION OF GLYCOSIDIC LINKAGES

### 6.1 Experimental approach

Methylation analysis has been an important method in structural analysis of polysaccharides for many years (14) and recently with the introduction of microtechniques for methylation (15) and subsequent analysis by g.l.c. and m.s. (16) the utility of the method has been greatly enhanced. Briefly, the method involves the complete methylation of a polysaccharide, the hydrolysis of the methylated product to a mixture of partially methylated monosaccharides, the reduction of the methylated monosaccharides, the acetylation of the reduced products and finally, analysis of the partially methylated alditol acetates by g.l.c. and m.s. The positions of the glycosidic linkages between monosaccharide units correspond to the positions of the free hydroxyl groups in the methylated monosaccharides. Methylation analysis does not give information on the anomeric configuration of the glycosidic linkages nor on the sequence of the monosaccharide residues in the polysaccharide. The latter determinations must be done by other methods and many ingeneous techniques have been devised for this purpose.

In methylation analysis it is essential that a complete methylation of all of the hydroxyl groups of a polysaccharide be achieved. A method has been developed using a strong base, methylsulphinyl methyl sodium, for ionizing free hydroxyl groups and methyl iodide for methylating these groups (see also Chapter 5, Section 7 and Chapter 6, Section 3.2). In the reaction alkoxide ions are generated readily from free hydroxyl groups by the strong base and methylation of these ions occurs rapidly. Dimethyl sulphoxide (DMSO) is used as the solvent for the reagents and the reactants. Methylsulphinyl sodium is a powerful base and effects the complete ionization of hydroxyl groups but the reaction is relatively slow and requires several hours for completion. The methyl iodide reacts rapidly with the alkoxyl ions resulting in the methylation of the alkoxide ion generated from the free hydroxyl groups in the polysaccharide. A reaction sequence for the methylation of a 4-substituted monosaccharide residue of a polysaccharide is shown in Equation 1.

$$(1)$$

A satisfactory method for measuring the completeness of the methylation is not available but incomplete methylation will be shown by the types of methylation products obtained or by i.r. measurements which will detect unsubstituted hydroxyl groups. Generally neutral polysaccharides are completely methylated by the method described in Section 6.4 and special precautions are not needed. Polysaccharides containing uronic acid or hexosamine residues are more difficult to methylate and may yield secondary products. For example, a uronic acid residue can yield a ketal derivative by reaction of the carboxyl group with the methylsulphinyl methyl sodium and the ketals may undergo elimination reactions yielding undesirable side products. Hexosamine residues with $N$ acetyl groups yield the $N$-methyl $N$-acetamido, $O$-methyl derivative of the hexo-amine on methylation. Such derivatives may require special analytical techniques for identification.

When the methylation is complete, the derivatized polysaccharide is hydrolysed to constituent monosaccharide units which are methylated at specific positions. These

65

derivatives are reduced and acetylated prior to analysis by g.l.c. and m.s. With some polysaccharides it may be necessary to use special techniques of hydrolysis since the derivatives are not soluble in the mineral acids. A preliminary short hydrolysis of the methylated polysaccharide in 90% formic acid is used and this treatment disrupts aggregates of the methylated polysaccharides yielding a product which is more easily hydrolysed in mineral acids.

The partially methylated monosaccharides obtained on acid hydrolysis are then converted to partially methylated alditol acetates or, in some cases to peracetylated aldononitriles prior to analysis. The alditol acetates are prepared by reduction of the methylated monosaccharides with sodium borohydride followed by acetylation with acetic anhydride in pyridine. The partially methylated alditol acetates are analysed and identified by g.l.c. and m.s. The preparation of aldononitriles involves reacting the methylated monosaccharides with hydroxylamine and then acetylating the product with acetic anhydride in pyridine. Analysis and identification of these derivatives is also done by g.l.c. and m.s.

## 6.2 Mass spectrometry

Methylated alditol acetates when subjected to a stream of high energy electrons at elevated temperature yield cleavage fragments which can be characterized by m/e values from mass spectra data. Extensive and complete studies have been conducted on the identification of the major fragments from partially methylated alditol acetates (16,17). The fragmentation of the acetates occurs in accord with a number of empirical rules formulated from experimental observations. First it should be pointed out that fragments which arise from methylated alditol acetates of stereoisomers with the same substitution patterns are identical in m/e values and in relative abundance. Such compounds are differentiated on the basis of retention times on g.l.c.

Second, it has been found that ion fragments produced from the methylated alditol acetates on electron impact are of two types, primary fragments which arise from an initial fission of a carbon-carbon bond and secondary fragments which arise by loss of some groups of atoms from the primary fragments. Primary fragments are produced in high abundance by fission of methylated alditol acetates between contiguous carbon atoms which carry methoxyl groups. The positive ion produced in the fragmentation is detected in the mass spectrometer. These ions are characterized by the mass to electron ratio (m/e). Negative ions are also produced but these are dissipated in the ionization chamber.

Illustrations of the fission patterns which yield primary fragments are shown in Chapter 1, Section 7.

## 6.3 Combined gas liquid chromatography and mass spectrometry

The analysis of the methylated alditol acetates is achieved in a unit in which a gas liquid chromatograph and a mass spectrometer are coupled for direct analysis of effluent from the column. Many types of instruments are available commercially for this purpose. A Varian 1400 Aerograph coupled to a DuPont 21-490 Mass Spectrometer is in use in this laboratory. Others types of instruments which can be used are the LKB combined gas chromatography-mass spectrometer and the Hewlitt Packard 5985 gas liquid chromatograph and mass spectrometer. The separation of the sugar derivatives by g.l.c. requires the use of suitable column packing materials, appropriate temperatures and proper flow rate for the carrier gas (18, see also Chapter 1, Section 6). The selection of the packing will be determined by the type of compounds and derivatives to be separated. The resolving power of the packing material for various derivatives is dependent on the type of organic liquid used to coat the inert support and the solubility of the derivatives in the phase. A column packing of 3% OV-225 on 80/110 Supelcoport gives good separations of partially methylated alditol acetates. Other column packings that can be used are ECNSS-M, OS-138, SP-2340 and SP-2330. The columns in use are made of stainless steel or of glass. With glass columns less derivative is adsorbed on the glass and such columns will give higher sensitivities. Analytical units in which gas liquid chromatographs and mass spectrometers are combined also require a gas stream splitter in order to divert some of the carrier gas and derivative into the mass spectrometer. The methodology of the g.l.c. and g.l.c.-m.s. is outlined in Chapter 1, Sections 6 and 7.

A polysaccharide of glucose and galactose from the cell wall of *S. faecalis* has been subjected to structural analysis by the above methylation technique and other methods (19). From these data, the structure for a repeating unit of the polysaccharide has been deduced as shown in *Formula 2*.

$$\rightarrow 4)\text{-}\beta\text{-}D\text{-}Glc}p\text{-}(1 \rightarrow 4) \ \beta \ D \ Glc}p\text{-}(1 \rightarrow 4)\text{-}\beta\text{-}D\text{-}Gal}p\text{-}(1 \rightarrow$$

$$6$$
$$\uparrow$$
$$1$$
$$\beta\text{-}D\text{-}Glc}p$$
$$4$$
$$\uparrow$$
$$1$$
$$\beta\text{-}D\text{-}Gal}p$$

**Formula 2**

The gas liquid chromatographic pattern for the methylated alditol acetates from this polysaccharide is reproduced in *Figure 4*. The derivatives responsible for the various peaks are labelled in the Figure and the peak heights indicate the relative amounts of the components.

## 6.4 **Methylation**

### 6.4.1 *Methylsulphinyl methyl sodium preparation*

Methylsulphinyl methyl sodium is prepared from DMSO and sodium hydride as follows (see also Chapter 5, Section 7 and Chapter 6, Section 3.2).

(i)     Thoroughly dry DMSO by stirring 500 ml of reagent grade DMSO with excess powdered calcium hydride for several hours at 65°C. Distil DMSO under nitrogen and at reduced pressure.

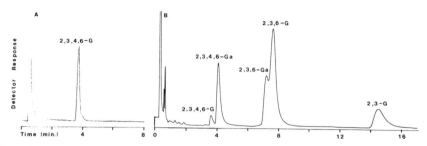

**Figure 4.** Gas liquid chromatography patterns of methylated alditol acetates of glucose (**A**) and of polysaccharides of glucose and galactose (**B**): 2,3,4,6-G = 1,5-di-O-acetyl-2,3,4,6-tetra-O-methyl glucitol, 2,3,4,6-Ga = 1,5-di-O-acetyl-2,3,4,6-tetra-O-methyl galacitol, 2,3,6-Ga = 1,4,5-tri-O-acetyl-2,3,6-tri-O-methyl galactitol, 2,3,6-G = 1,4,5-tri-O-acetyl-2,3,6-tri-O-methyl glucitol and 2,3-G = 1,4,5,6-tetra-O-acetyl-2,3-di-O-methyl glucitol (24).

(ii) Transfer the DMSO into brown bottles stoppered tightly with serum caps. Store over molecular sieves in the cold room.

(iii) For use, thaw the frozen solvent and remove a sample with a dry glass syringe. Transfer into the desired vessel.

(iv) Prepare the methylsulphinyl methyl sodium by adding 30 ml of dried DMSO into a three-necked 250-ml flask containing 1.5 g of 57% suspension of sodium hydride in mineral oil mixed with 10 ml of anhydrous ether.

(v) Evacuate the flask and flush with nitrogen three times under vacuum. Add the DMSO from a dry syringe through the serum cap.

(vi) Attach a glass syringe to the flask through the rubber cap to facilitate the escape of hydrogen.

(vii) Sonicate the reaction mixture in the flask in a water bath of 50°C for 2−3 h. The synthesis of methylsulphinyl methyl sodium is complete when a transparent green colour appears in the solution in the flask.

(viii) Transfer the dimethylsulphinyl methyl sodium to several 5-ml serum bottles which are capped with serum caps and flushed with nitrogen. Store in the frozen state at 4°C in the dark.

(ix) For methylations, thaw the methylsulphinyl methyl sodium and remove the sample through the rubber cap by means of a dry glass syringe. Add the solution directly from the syringe into the flask containing the polysaccharide.

### 6.4.2 *Methylation procedure*

A satisfactory procedure used for methylations in this laboratory (7) is as follows.

(i) Place a sample of 2−5 mg of the thoroughly dry polysaccharide into a dry 25 ml round bottom flask.

(ii) Fit the flask with a rubber serum cap and evacuate for 1 h to further dry the polysaccharide.

(iii) Continue the evacuation and flush the flask with nitrogen three times.

(iv) Add 1 ml of dry DMSO from a dry glass syringe and sonicate the reaction mixture for 20 min at room temperature.

(v) Add 0.4 ml of methylsulphinyl methyl sodium solution using a dry glass syringe.

(vi) Liquify the resulting gel by sonication for 20 min at 25°C.

(vii)   Add 0.3 ml of dry methyl iodide and sonicate the mixture for 15 min at 25°C.

(viii)  Maintain the reactants at room temperature for an additional 1–6 h.

(ix)    Remove the serum cap from the flask and add a small amount of water to neutralize excess methylsulphinyl methyl sodium.

(x)     Transfer the resulting suspension into 12-ml clinical centrifuge tubes and extract the methylated polysaccharide in the suspension with 4 ml of chloroform.

(xi)    Wash the chloroform layer with 3 ml of water and centrifuge at high speed in a clinical centrifuge. Remove and discard the top water layer and wash the chloroform layer three more times with 3 ml aliquots of water discarding the water layer after each washing.

(xii)   After the last washing dry the inside of the tubes with a Kimwipe and add anhydrous magnesium sulphate to remove the last traces of water.

(xiii)  Remove the magnesium sulphate by filtration through glass filter in a Pasteur pipette.

(xiv)   Collect the chloroform filtrate with the methylated polysaccharides in a 10-ml ampoule.

(xv)    Evaporate the chloroform under a stream of nitrogen at 40°C until approximately 1 ml of solution of the methylated polysaccharide remains.

(xvi)   Purify the derivative by chromatography on a column of 2 g of Sephadex LH-20 which has been washed with a solvent of chloroform acetone mixture (2:1).

(xvii)  Place the sample of methylated polysaccharide on the column and elute the derivative with the chloforom-acetone mixture (2:1 by vol.). The methylated polysaccharide can be detected in the Sephadex as a light pink band.

(xviii) Collect this band and transfer to a 10-ml ampoule.

(xix)   Remove the solvent by evaporation under a stream of nitrogen at 40°C.

## 6.4.3 *Hydrolysis procedure*

(i)     Suspend the methylated polysaccharide in the 10 ml ampoule in 1 ml of 90% formic acid and sonicate for 15 min at room temperature.

(ii)    Flush the ampoule with nitrogen, then seal by heat. Heat in an oven at 105°C for 1.5 h.

(iii)   Open the ampoule and evaporate the formic acid by heating to 40°C.

(iv)    Add 2 ml of 0.15 M $H_2SO_4$ to the ampoule containing the dry methylated polysaccharide.

(v)     Flush the ampoule with nitrogen, reseal and reheat for 12–18 h at 105°C in an oven.

(vi)    Open the ampoule and add barium carbonate until the acid is neutralized.

(vii)   Add 0.1 M NaOH to aid in flocculation of the precipitate.

(viii)  Transfer the mixture into 12-ml clinical centrifuge tubes and centrifuge.

(ix)    Recover the supernatant by filtration into 25-ml Erlenmeyer flasks.

## 6.4.4 *Reduction*

(i)     Reduce partially methylated monosaccharides obtained by hydrolysis of the methylated polysaccharide with 1 mg of sodium borohydride for 12 h at room temperature.

(ii)   Neutralize the mixture by acidification with Dowex 50 ($H^+$) ion exchange resin and evaporate the solvent in a rotary evaporator at a temperature below 40°C and at reduced pressure.

(iii)  Dissolve the residue in redistilled methanol and transfer into a 10-ml ampoule.

(iv)   Remove the methanol by heating at 40°C in a water bath and with a stream of nitrogen.

(v)    Add methanol and evaporate three additional times in order to insure that all the borate is removed as volatile trimethyl borate.

### 6.4.5 *Acetylation*

(i)    Acetylate the methylated, hydrolysed and reduced polysaccharide products with acetic anhydride in pyridine in a 10 ml ampoule.

(ii)   Add 0.5 ml of equal parts of acetic anhydride and pyridine, flush the ampoule with nitrogen, seal and heat at 105°C for 2 h.

(iii)  Open the ampoule and evaporate the acetic anhydride and pyridine in a stream of nitrogen at 40°C.

(iv)   Add ether to the reaction mixture and evaporate the solvent.

(v)    Repeat this treatment several times to remove all traces of acetic anhydride and pyridine.

(vi)   Finally dissolve the sample in a small amount of chloroform and use aliquots of the chloroform solution for analysis by g.l.c. and m.s.

### 6.4.6 *Methylated peracetyl aldononitriles*

The derivatives of the methylated polysaccharides may be analysed as the aldononitrile derivatives (20).

(i)    React the methylated monosaccharides with hydroxylamine hydrochloride in a pyridine solvent at 60°C for 20 min with vigorous stirring.

(ii)   Add acetic anhydride and heat the reaction mixture for an additional 20 min at 60°C.

(iii)  Extract the methylated peracetyl aldononitriles with chloroform and use appropriate aliquots for analysis by g.l.c. and m.s.

(iv)   Analyse the aldononitrile acetates on a column of 5% butanediol succinate on Supelcoport 80/100.

(v)    The mass spectra data for the methylated peracetylated aldononitrile derivatives of monosaccharides are obtained in a similar manner as with the methylated alditol acetates.

The m/e values for the fragments from these derivatives containing the nitrile group will be quite different from those of the corresponding alditol acetates. The unsymmetrical nature of the aldononitriles aids in the differentiation of the monosaccharides. Both the nitrile and the alditol derivatives possess the advantage that only one isomer of the saccharide is produced and hence overlapping of peaks on gas liquid chromatography is not likely. The retention times and the nature of the mass fragments from peracetyl aldononitriles are still under compilation.

# 7. DETERMINATION OF THE SEQUENCE OF MONOSACCHARIDE RESIDUES

## 7.1 Methylation analysis

The details of the procedure for the methylation of polysaccharides and the analysis of the products have been discussed in Section 6.1. From the chromatographic and mass spectra data, the types of residues and the positions of the glycosidic bonds can be deduced. However the anomeric configuration of the bonds and the sequence of residues in the polysaccharide cannot be established from these data alone. Additional types of analyses are needed in order to establish the absolute sequence of monosaccharide residues in a polysaccharide.

## 7.2 Periodate oxidation

Periodate oxidation of polysaccharides is a valuable analytical technique that has been used in the structural characterization and monosaccharide sequence determination of polysaccharides (21). The reactions of periodate oxidation of monosaccharides and oligosaccharides have been discussed in Chapter 2 and similar types of reactions occur during the oxidation of polysaccharides. Briefly, carbohydrate residues containing glycol groups are oxidized to dialdehydes but if residues contain hydroxyl groups on three adjacent carbons formic acid can also be produced. In practice, the oxidized residues are reduced with sodium borohydride yielding acetal groups from some of the aldehyde groups. Additional information on structure can be obtained by removing the acetal groups with mild acid and identifying the residual fragments. The oxidized and reduced polysaccharide can also be subjected to a second periodate oxidation and additional structural information can be obtained. The reactions for the oxidation, the reduction and the hydrolysis of a 4-substituted residue are illustrated in Equation 2 in which the $R_1$ and $R_2$ are glycosyl residues substituted in a manner that periodate oxidation does not occur.

It was observed in early studies that the periodate oxidation of monosaccharides resulted in the formation of intra hemiacetal bonds between the aldehydic groups generated by the oxidation and unsubstituted hydroxyl groups of the monosaccharides. The formation of hemiacetal bonds also occurs during the oxidation of polysaccharides

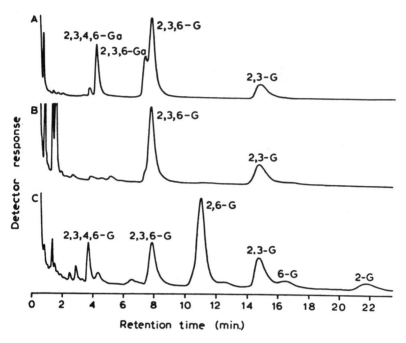

**Figure 5.** Gas-liquid chromatographic patterns for the partially methylated alditol acetates from the native (**A**), the periodate-oxidized and borohydride-reduced (**B**), and the periodate oxidized (**C**) polysaccharide of glucose and galactose: 2,3,4,6-Ga = 1,5-di-*O*-acetyl-2,3,4,6-tetra-*O*-methyl galactitol, 2,3,4,6-G = 1,5-di-*O*-acetyl-2,3,4,6-tetra-*O*-methyl glucitol, 2,3,6-Ga = 1,4,5-tri-*O*-acetyl-2,3,6-tri-*O*-methyl galactitol, 2,3,6-G = 1,4,5,-tri-*O*-acetyl-2,3,6-tri-*O*-methyl glucitol, 2,3-G = 1,4,5,6-tetra-*O*-acetyl-2,3-di-*O*-methyl glucitol, 2,6-G = 1,3,4,5-tetra-*O*-acetyl-2,6-di-*O*-methyl glucitol, 6-G = 1,2,3,4,5-penta-*O*-acetyl-6-di-*O*-methyl glucitol, and 2-G = 1,3,4,5,6-penta-*O*-acetyl-2-*O*-methyl glucitol (22).

(22). *Figure 5* shows results of methylation analysis of the native and periodate oxidized diheteropolysaccharide of glucose and galactose from *S. faecalis* with a repeating unit of the structure shown in *Formula 2*. Frame A of *Figure 5* shows the methylated products from the native polysaccharide, Frame B, the methylated products from the periodate oxidized and borohydride reduced polysaccharide, and Frame C, the methylated products from the periodate oxidized polysaccharide. It will be noted that a large amount of 2,6-di-*O*-methyl glucose and small amounts of 2-*O*-methyl and 6-*O*-methyl glucose are present in the pattern in Frame C. Such products show that the hydroxyl groups at positions 3, 6 and 2 of some unoxidized glucose residues were not methylated but had reacted with the aldehyde groups of oxidized galactose residues to yield hemiacetal bonds. The formation of hemiacetal bonds in periodate oxidized polysaccharides occurs by the reaction sequence shown in Equation 3.

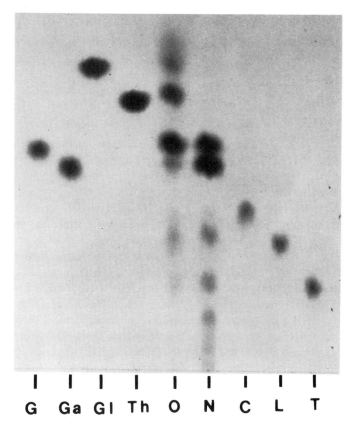

**Figure 6.** A chromatogram of acid hydrolysates of periodate-oxidized and borohydride reduced (O) and native (N) polysaccharide of glucose and galactose: G = glucose, Ga = galactose, Gl = glycerol, Th = threitol, C = cellobiose, L = lactose, T = trisaccharide of glucose with $\alpha$-$(1\rightarrow4)$ and $\alpha$-$(1\rightarrow6)$ linkages (24).

To obtain the theoretical amount of oxidation of a polysaccharide it is generally necessary to reduce the aldehydic groups with sodium borohydride or other reducing agents and to re-oxidize the product with periodate. One such analytical scheme has become known as the Smith degradation (21,23). The oxidation, reduction and mild acid hydrolysis are used to obtain products which can be identified and utilized in structural elucidations. Paper chromatography can be used to identify products and *Figure 6* shows products from the periodate oxidized and borohydride reduced polysaccharide of glucose and galactose (24). It should be noted that glycerol, threitol, glucose, and some oligosaccharides were produced in large amounts and galactose and other oligosaccharides in small amounts on hydrolysis of the oxidized polysaccharide. For comparative purposes a chromatogram of a partial acid hydrolysate of the native polysaccharide is also shown in *Figure 6*. In the hydrolysate of the oxidized and reduced polysaccharide, the glycerol arises from terminal galactose units of the oxidized polysaccharide and the threitol arises from 4-substituted internal galactose units which are linked by $(1\rightarrow4)$ glycosidic linkages to other residues of the polysaccharides. Polysaccharides with other glycosidic bonds as for example $(1\rightarrow2)$, $(1\rightarrow3)$, $(1\rightarrow4)$ and $(1\rightarrow6)$ linkages will yield

**Formulae 3, 4 and 5**

different products. The structures of some of these reaction products are shown in *Formulae 3, 4* and *5*. These products would arise from 2-substituted, 4-substituted and 6-substituted glycosyl residues, respectively. The 3-substituted residues would not be susceptible to oxidation.

Another analytical scheme for identifying periodate oxidation products of polysaccharides is known as the Barry degradation and has been described in a recent review (23). In the Barry degradation the periodate oxidized polysaccharide is reacted with phenylhydrazine in dilute acetic acid. Phenylhydrazones are formed with the aldehyde groups generated by the oxidation. The phenylhydrazone containing units are released by hydrolysis. Structural data from Barry degradations have been obtained for gum arabic, galactan, arabinogalactan and nigeran. Some polysaccharides can be degraded by the Barry method with a sequential removal of monosaccharide residues. Then the sequence of carbohydrate residues in the polysaccharide chain can be determined. An application of the latter method was in the determination of the sequence of an oligosaccharide isolated from a periodate-oxidized and borohydride-reduced oligosaccharide from a pnueomococcal polysaccharide. The sequence of monosaccharides for the repeating unit of this polysaccharide is shown in *Formula 6*.

$$\alpha\text{-L-Ara}f\text{-}(1\rightarrow3)\text{-}\alpha\text{-D-Glc}p\text{-}(1\rightarrow2)\text{-}\alpha\text{-L-Ara}f\text{-}(1\rightarrow3)\text{-}\alpha\text{-D-Gal}p\text{-}(1\rightarrow2)\text{-}$$

**Formula 6**

A laboratory protocol for the periodate oxidation of a polysaccharide with the repeating unit shown in *Formula 2* is as follows.

(i) Dissolve 50 mg of the polysaccharide in 50 ml of 0.02 M sodium periodate, adjusted to pH 4.5.

(ii) Cover the reaction flask with aluminum foil and place the flask and reaction mixture in a cold room at 4°C for 18 h. At the end of this time destroy the excess periodate by addition of a few ml of ethylene glycol.

(iii) Remove the low molecular weight substances by dialysis against distilled water for 48 h.

(iv) Lyophilize the oxidized polysacharide to dryness.

(v) Dissolve 20 mg of the periodate-oxidized polysaccharide in 10 ml of water.

(vi) Add 5 mg of sodium borohydride and maintain the reaction mixture for 24 h at room temperature.

(vii) Dialyse the reaction mixture against distilled water for 24 h.

(viii) Lyophilize the sample to dryness.

(ix) Dissolve 10 mg of the periodate oxidized and borohydride reduced polysaccharide in 0.5 ml of 0.02 M HCl in a small tube, stopper the tube and heat the reaction mixture in a boiling water bath for 20 min.

(x)     Hydrolyse 2 mg of the native polysaccharide in 0.1 ml of 0.1 M HCl for 3 h in a boiling water bath.

(xi)    Analyse both hydrolysates for reducing sugars by paper chromatography using two ascents of the solvent system of *n*-butyl alcohol, pyridine and water (6:4:3 by vol.) as described in Chapter 1, Section 4.

(xii)   Stain the finished chromatogram with silver nitrate and sodium hydroxide reagents.

A photograph of the chromatogram is reproduced in *Figure 6* (24). It can be seen that the hydrolysate of the oxidized and reduced polysaccharide contained high amounts of glycerol (Rf 0.76), threitol (Rf 0.67), glucose (Rf 0.55) and a trisaccharide (Rf 0.25) and trace amounts of galactose (Rf 0.50), a disaccharide (Rf 0.36) and a tetrasaccharide (Rf 0.19). The hydrolysate of the native polysaccharide contained glucose, galactose and a series of reducing oligosaccharides. As stated above, the hydrolysate of the oxidized and reduced polysaccharide contained a large amount of a compound which migrated on paper chromatograms with an Rf value typical of trisaccharides. The compound was isolated by preparative paper chromatography and found to be composed of glucose. Methylation analysis showed that the methyl alditol acetates obtained were 1,5-di-*O*-acetyl-2,3,4,6-tetra-*O*-methyl hexitol and 1,4,5,6-tetra-*O*-acetyl-2,3-di-*O*-methyl hexitol in the molar ratio of 2:1. The structure of this trisaccharide is a $\alpha$-D-glucopyranosyl-$(1\rightarrow4)$-[$\alpha$-D-glucopyranosyl-$(1\rightarrow6)$]-$\alpha$-D-glucopyranose. This information was essential for the elucidation of the complete molecular structure of the polysaccharide.

## 7.3 Acetolysis

Acetolysis of polysaccharides results in the complete acetylation of the free hydroxyl groups of the polysaccharide and the selective cleavage of glycosidic bonds. Rate constants for the hydrolysis of glycosidic bonds of representative disaccharides have been determined and these values are recorded in *Table 3* (23). In view of the differences in the rate constants for the cleavage of glycosidic bonds under acetolysis conditions it is possible to fragment polysaccharides preferentially at specific glycosidic bonds. It is noted in *Table 3* that the $(1\rightarrow6)$ glycosidic linkage is highly susceptible to cleavage by acetolysis whereas the $(1\rightarrow2)$ and the $(1\rightarrow3)$ linkages are comparatively resistant.

Acetolysis reactions are performed by heating the polysaccharides in a mixture of acetic anhydride, acetic acid and sulphuric acid generally in a ratio of 10:10:1. In this mixture, the polysaccharide is first acetylated and then hydrolysed by cleavage of certain of the glycosidic bonds. Studies on the mechanism of the acetolysis reactions have been conducted with oligosaccharides. It has been found that acetolysis is retarded by the presence of electronegative groups in the aglycon portion of the molecule and by the type of anomeric configuration of the bonds. Compounds with glycosidic linkages in the $\alpha$ configuration are more susceptible to acetolysis than compounds with $\beta$-glycosidic linkages. The positions of the glycosidic linkages in the oligosaccharide also affect acetolysis rates.

On the basis of these results, a mechanism proposed for acetolysis reactions involves

**Table 3.** Rates of acetolysis of disaccharides (23).

| Disaccharide | Relative rate[a] |
|---|---|
| $\alpha$-D-Glc-(1→2)-D-Glc | 0.18 |
| $\alpha$-D-Glc-(1→3)-D-Glc | 0.27 |
| $\alpha$-D-Glc-(1→4)-D-Glc | 1.0 |
| $\alpha$-D-Glc-(1→6)-D-Glc | 7.9 |
| $\beta$-D-Glc-(1→2)-D-Glc | 1.4 |
| $\beta$-D-Glc-(1→3)-D-Glc | 1.9 |
| $\beta$-D-Glc-(1→4)-D-Glc | 1.2 |
| $\beta$-D-Glc-(1→6)-D-Glc | 58 |
| $\alpha$-D-Man-(1→2)-D-Man | 0.21 |
| $\beta$-D-Man-(1→3)-D-Man | 2.9 |
| $\alpha$-D-Man-(1→4)-D-Man | 1.4 |
| $\alpha$-D-Man-(1→6)-D-Man | 60 |
| $\beta$-D-Gal-(1→4)-D-Gal | 4.1 |
| $\alpha$-D-Gal-(1→6)-D-Gal | 41 |
| $\alpha$-D-Glc-(1→3)-D-Fru | 71 |

[a]Rate of acetolysis, relative to that of maltose.

the formation of an acyclic intermediate followed by cleavage of glycosidic bonds and release of the acetolysis fragments as shown in Equation 4 (23).

$$(4)$$

Acetolysis of polysaccharides can be used as a complementary method to conventional acid hydrolysis. The two procedures will yield different fragments from a polysaccharide and such fragments can be used to obtain structural information. The (1→6) glycosidic bond in polysaccharides is relatively stable to acid hydrolysis but such bonds are much more readily cleaved by acetolysis.

The acetolysis method has been especially useful for investigating the structure of polysaccharides from microorganisms. The method has been used in an extensive series of studies designed to elucidate the structure of yeast mannans (25). The acetolysis was performed by the acetic anhydride-acetic acid sulphuric acid method and the analysis for oligosaccharide fragments was performed by chromatography on Sepharose G-25. The yeast mannan contains (1→6), (1→2) and (1→3) linkages and acetolysis experiments

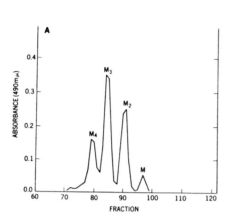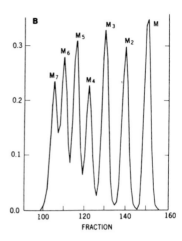

**Figure 7.** Filtration patterns of the manno-oligosaccharides on Sephadex G-15: **Frame A** an acetolysis of yeast mannan; **Frame B**, synthetic mixture of manno-oligosaccharides (25).

revealed the nature and the distribution of the oligossaccharide side chains on the main chain of the mannan. The results of one acetolysis experiment in which a deacetylation of oligosaccharides was performed prior to Bio-Gel chromatography are shown in Frame A of *Figure 7*. The yeast mannan used in this study was found to yield oligosaccharides of 2, 3 or 4 mannose units. Frame B of *Figure 7* shows the resolution of reference mannose oligosaccharides by Bio-Gel chromatography. These data are used to identify products of acetolysis of mannans.

In *Figure 7*, Frame A, it will be noted that the amount of the trisaccharide was apparently twice that of the other oligosaccharides. It was suggested that the trisaccharide fraction was composed of isomeric forms. Methylation studies indeed confirmed this suggestion. The repeating unit proposed for the structure of the mannan is shown in *Formula 7*.

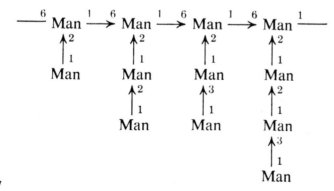

**Formula 7**

The acetolysis procedure has been applied in the analysis of the structure of a bacterial polysaccharide of glucose and galactose. The protocol for this experiment was as follows.

(i)    Dissolve 10 mg of the polysaccharide in 1 ml of the acetolysis solution of acetic anhydride, glacial acetic acid and concentrated $H_2SO_4$ (10:10:1 by vol.).

(ii)    Heat the solution in a stoppered reaction vessel for 3 h at 40°C.

(iii)   Add pyridine to the mixture to stop the reaction and evaporate the solvents in a stream of nitrogen at 40°C.

(iv)    Deacetylate the products by dissolving the sample in methanol containing a catalytic amount of barium methoxide. Maintain the reaction mixture at room temperature for 15 min.

(v)     Neutralize the base by adding dry ice and then evaporate the solvent.

(vi)    Suspend the residues in a small amount of water and remove the barium carbonate by centrifugation.

(vii)   Concentrate the supernatant to a small volume (0.5 – 1 ml) and analyse for carbohydrate types by paper chromatography in a solvent of *n*-butyl alcohol-pyridine and water (6:4:3 by vol.). Two reducing products are detected on the paper chromatogram, one which moves on paper chromatography with an Rf value of 0.68 and typical of disaccharides and the other a high molecular weight product which remains at the origin.

(viii)  Isolate the two carbohydrate products by preparative paper chromatography and subject the compounds to methylation analysis.

The methylated alditol acetates which were identified from the oligosaccharides were 1,5-di-*O*-acetyl 2,3,4,6-tetra-*O*-methyl galactitol and 1,4,5-tri-*O*-acetyl-2,3,6-tri-*O*-methyl glucitol in the molar ratio of 1:1. Both compounds are susceptible to hydrolysis by $\beta$-galactosidase. In view of the above results the oligosaccharide is lactose and the polysaccharide fragment is also a $\beta$-linked polymer. The acetolysis results show that the lactose units form the side chains of the polysaccharide and are linked by $(1 \rightarrow 6)$ bonds to the main chain of the polysaccharide. The methylated alditol acetates from the high molecular weight product were 1,4,5-tri-*O*-acetyl-2,3,6-tri-*O*-methyl galactitol and 1,4,5-tri-*O*-acetyl-2,3,6-tri-*O*-methyl glucitol in the ratio of 1:2. This product is derived from the main chain of the polysaccharide since cleavage of the main glycosidic bonds of the main chain did not occur by acetolysis reactions.

## 7.4 Methanolysis

Simultaneous hydrolysis and methanolysis of polysaccharides to yield methyl glycosides can be effected at elevated temperatures in anhydrous methanolic hydrogen chloride (26). On prolonged heating the polysaccharides are converted to the methyl glycosides of the constituent monosaccharides but on short heating methyl glycosides of oligosaccharides will be produced. The reaction is carried out at temperatures in the range of 80 – 100°C and for periods of 1 – 5 h.

(i)     Dissolve 0.5 mg of the polysaccharide in 1 ml of anhydrous methanolic hydrogen chloride and heat at 95°C for 5 h.

(ii)    Neutralize the reaction mixture with powdered silver carbonate and centrifuge the suspension to remove the precipitated silver salt.

(iii)   Transfer the supernatant to a vial and remove the solvent with a stream of nitrogen.

(iv)    The residue consists of a mixture of carbohydrate derivatives from the polysaccharide and is used in subsequent analyses.

(v)     Reacetylation of the amino groups of the polyaccharide is effected if necessary in acetic anhydride and pyridine at room temperature for 6 h.

The methyl glycosides can be converted to the trimethylsilyl derivatives for analyses or to the acetyl derivatives. Both types of derivatives can be analysed by g.l.c. with the appropriate column packing. An alternative method for use on a micro scale is described in Chapter 1, Section 6.2.1. The trimethylsilyl or the acetyl derivative can be prepared from the reduced monosaccharides by first removing the methyl group by acid hydrolysis and then reducing the aldehyde group with sodium borohydride. The derivatives are prepared using appropriate silylating or acetylating reagents. Derivatives of the reduced monosaccharides are advantageous to use since $\alpha$- and $\beta$-anomers cannot be formed because of the destruction of the asymmetry at carbon 1 by the reductions. As a result one peak is obtained for individual compounds on g.l.c. analysis. Quantitative values for the monosaccharides will be obtained by integrating the areas under the peaks of the patterns and comparing the areas with those for standards of the monosaccharides.

## 7.5 Hydrolysis by acids

The rate constants for the hydrolysis of the glycosidic bonds in polysaccharides vary greatly and are affected by factors such as the positions of glycosidic linkages, the ring structure of the residues and the anomeric configuration of the linkages (23). It has been observed that compounds with $(1 \rightarrow 2)$ linkages are more easily hydrolysed than compounds with $(1 \leftarrow 6)$ linkages and compounds with the $\alpha$-anomeric configuration are more easily hydrolysed than compounds with the $\beta$-configuration. The types of the monosaccharide residues constituting the polymer also affect the rate constants for the hydrolysis of glycosidic bonds. Thus linkages between uronic acid residues and amino sugar residues are more slowly hydrolysed than similar linkages between other residues. The hydrolysis of polysaccharides containing the uronic acid or amino sugar residues may require the use of special techniques. A pre-treatment of the polysaccharide with an organic acid is often used. It may be necessary to modify uronic acid or amino sugar residues by chemical reactions and the modified residues can be removed more readily. In addition to promoting hydrolysis of the polysaccharide, the modification can yield information on the structure of the polysaccharide. The reduction of the carboxyl group of uronic acid not only facilitates hydrolysis but also can be used to identify the type of uronic acid and the type of the glycosidic bonds. Results of one such experiment are presented in *Figure 8* for the polysaccharide with a repeating unit presented in *For-*

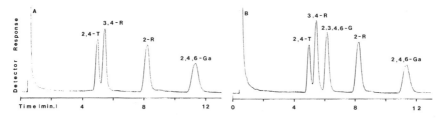

**Figure 8.** Gas liquid chromatography patterns for the methylated alditol acetates from the native polysaccharide from *S. bovis* (**Frame A**) and the polysaccharide after reduction of the D-glucuronic acid residues (**Frame B**): 2,4,T = 1,3,5,-tri-*O*-acetyl-2,4-di-*O*-methyl-6-deoxy talitol; 3,4-R = 1,2,5-tri-*O*-acetyl-3,4-di-*O*-methyl rhamnitol; 2-R = 1,3,4,5-tetra-*O*-acetyl-2-*O*-methyl rhamnitol; 2,4,6-Ga = 1,3,5-tri-*O*-acetyl-2,4,6-tri-*O*-methyl galactitol and 2,3,4,6-G = 1,5-diacetyl-2,3,4,6-tetra-*O*-methyl glucitol (7).

*mula 1.* Following reduction of the polysaccharide and methylation analysis, a tetramethyl glucose derivative was identified by g.l.c. as shown by the data in *Figure 8.* In order to obtain this derivative the glucuronic acid residues must be present as terminal units in the original polysaccharide. Modification of the amino sugar residues by removal of the acetyl group by the trifluoroacetic acid and other methods has been employed to convert such polysaccharides into more readily hydrolysable form.

Partial acid hydrolysis by acids or enzymes has been an extremely useful technique in structural analysis of polysaccharides. By partial acid hydrolysis, polysaccharides can be cleaved randomly into a mixture of oligosaccharides or may be cleaved preferentially at a specific bond to give a high yield of products with uniform structure. The oligosaccharides are isolated and then subjected to appropriate analysis. If a sufficient number of oligosaccharides are prepared which contain overlapping segments of the polysaccharide, the sequence of residues in the polysaccharide can be deduced.

Monosaccharides with furanose ring structures are very acid labile and such residues can be removed preferentially by mild acid hydrolysis. Analysis of the residual fragment can yield valuable structural data. Terminally linked residues are also hydrolysed

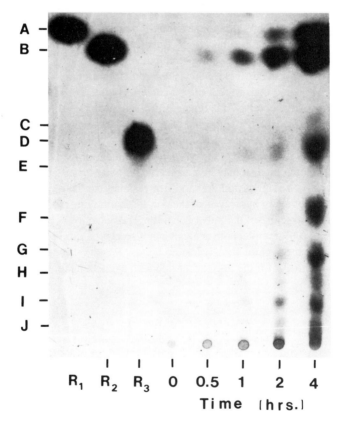

**Figure 9.** A paper chromatogram of reference compounds and hydrolysates of the polysaccharide of glucose and galactose as a function of time: $R_1$, $R_2$ and $R_3$ = reference compounds (glucose, galactose and lactose), A = glucose, B = galactose, C = cellobiose, D = lactose, E = galactose-($\beta$-1→6)-glucose, F = trisaccharide, G = tetrasaccharide, H = pentasaccharide, I = hexasaccharide and J = heptasaccharide (19).

preferentially. A polysaccharide of glucose and galactose was subjected to partial acid hydrolysis and the hydrolysate was analysed as a function of time by paper chromatographic methods. A photograph of the chromatogram is shown in *Figure 9*. The glycosidic linkages in this polysaccharide are hydrolysed at approximately the same rates and as a result a series of oligosaccharides and glucose and galactose are produced. It will be noted in the figure that in short periods of hydrolysis galactose was liberated indicating that the galactose units were terminal units. The oligosaccharides were isolated by preparative paper chromatography and subjected to further structural analysis.

The partial acid hydrolysis of methylated polysaccharides can also be effected to yield fragments valuable in structural analysis (27). One such hydrolysis was performed by heating the completely methylated polysaccharide at 90°C for 40 min in 2 M trifluoroacetic acid. Another type of hydrolysis was achieved by heating in 90% formic acid. In methylated polysaccharides the glycosidic bonds of polysaccharide derivatives are hydrolysed at different rates. As a result some of the linkages may be over-hydrolysed while others may be under-hydrolysed. Provided the differences in hydrolysis are not extreme, a mixture of derivatized oligosaccharides will be obtained representing all possible types of oligosaccharides. The derivatized oligosaccharides are reduced with borohydride and then separated by h.p.l.c. on suitable column packing. The elution of derivatized oligosaccharides from the column was effected with a solvent mixture of acetonitrile and water. Fractions of the eluate were collected in a fraction collector and then subjected to analysis by g.l.c. and m.s. for the determination of the monosaccharide sequence. The sequence of residues in the polysaccharide can then be deduced from the sequences of the oligosaccharides.

## 7.6 Hydrolysis with enzymes

The hydrolysis of polysaccharides can also be achieved with appropriate hydrolytic enzymes. Such hydrolysates will contain a series of oligosaccharides which can be isolated and used in structural studies. Enzymes which catalyse the hydrolysis of polysaccharides are of two types, the exohydrolases which hydrolyse terminal glycosidic linkages and liberate terminal monosaccharide units and the endohydrolases which hydrolyse internal glycosidic bonds at a specific position or randomly. The endohydrolases generally effect a progressive shortening of the complete polysaccharide with long segments of residues being produced in the initial stages of hydrolysis and short oligosaccharide segments in later stages. Some endohydrolases possess a specificity for a particular type of linkage and as a result a residual fragment is of relatively uniform structure. Valuable information on the structure and residue sequences of the fragments and on the polysaccharides will be obtained. For the hydrolysis of heteropolysaccharides the ideal situation is to have a series of exoglycosidases available, each of which is specific for a single type of terminal saccharide residue and each capable of removing such residues quantitatively. Such a series of enzymes could be employed sequentially for liberating terminal residues that would be identified at each step. On complete hydrolysis of a polysaccharide by exohydrolases and identification of the monosaccharides liberated, the sequence of the polysaccharide would be established. It has not yet been possible to obtain exoglycosidases needed for the complete hydrolysis of a polysaccharide. However it has been possible to prepare those enzymes needed for sequencing the oligosaccharide from a glycoprotein (28). The glycoprotein was of mammalian origin and the monosaccharide sequence of the main chain was determined to be Gal-GlcNAc-Man-Man-GlcNAc-GlcNAc- and of a side chain to be Gal-GlcNAc-Man- which is attached to the first mannose from the non-reducing end of the oligosaccharide. The enzymes employed for the sequential hydrolysis were $\beta$-galactosidase, *N*-acetyl-$\beta$-glucosaminidase, $\beta$-mannosidase and $\alpha$-mannosidase and the identification of the monosaccharides was by chromatography on a Bio-Gel P-4 column.

An enzymic procedure has been used for investigating the structure of an antigenic polysaccharide of glucose and galactose (19).

(i)     Dissolve 60 mg of the polysaccharide in 1.5 ml of water and mix with 0.75 ml of 4% almond $\beta$-glycosidase solution in 0.1 M phosphate buffer (pH 6.8). $\beta$-Glucosidase and $\beta$-galactosidase activities were present in the enzyme preparation and these enzymes effect the hydrolysis of glycosidic bonds of the polysaccharide.

(ii)    Maintain the enzymic digest at room temperature for a period of 44 h.

(iii)   Examine the digest periodically for the liberation of reducing sugars by paper chromatography. The $\beta$-glycosidases of almonds are exohydrolases and remove residues from the non-reducing ends of the polysaccharide with D-galactose being liberated early in the reaction and in highest amounts. On prolonged hydrolysis a small amount of glucose is also produced.

(iv)    At the end of 44 h inactivate the glycosidases with heat and precipitate the enzymes with an equal volume of 10% trichloroacetic acid.

(v)     Remove the precipitate by centrifugation, collect the supernatant and dialyse for 48 h against distilled water.

(vi)    Dry the enzyme-treated polysaccharide by lyophilization.

(vii)   Test the preparation by qualitative precipitin tests with homologous antiserum.

(viii)  Repeat the enzyme treatment if the polysaccharide still possesses antigenicity.

(ix)    After the second enzyme treatment, re-isolate the modified polysaccharide.

(x)     Subject 2 mg of the enzyme-treated polysaccharide to methylation analysis.

(xi)    Identify and quantitate the products by g.l.c. and m.s.

The quantitative data calculated for the native and enzymically modified polysaccharide are recorded in *Table 4*.

The enzymic method can be coupled with acid hydrolysis techniques (19) and additional structural information can be obtained.

**Table 4.** Moles of methylated monosaccharides from the native and enzymically modified polysaccharide of glucose and galactose on methylation analysis (19).

| Compound | Native | Modified | Difference |
|----------|--------|----------|------------|
| 2,3,4,6-Tetramethylglucose | 4.0 | 15.9 | +11.9 |
| 2,3,4,6-Tetramethylgalactose | 17.5 | 5.0 | −12.5 |
| 2,3,6-Trimethylgalactose | 17.6 | 17.8 | + 0.2 |
| 2,3,6-Trimethylglucose | 35.0 | 21.9 | −13.1 |
| 2,3-Dimethylglucose | 17.1 | 16.8 | −0.3 |

## 7.7 **Degradation reactions**

### 7.7.1 *Alkaline degradation*

The degradative action of alkalis on polysaccharides has been investigated for many years but the exact nature of the degradative reactions has not yet been determined (23). Variations in the types of degradation products obtained from polysaccharides with different structures have been noted. It is reasonably well established that alkaline degradation begins at the residue containing the reducing group and this residue is ultimately eliminated. In concentrated alkali, residues with free hydroxyl groups are also degraded by oxidation of the hydroxyl groups. The presence of substituents on hydroxyl groups of such residues markedly affects the degradative reactions and in some cases arrests the degradation process. This observation has been used to devise a method for determining the position of branch points of branched polysaccharides.

In dilute solutions of alkali and in the absence of oxygen, degradation of polysaccharides is not extensive but in more concentrated alkali and in the presence of oxygen profound changes occur in the polysaccharide. The initial reaction of the degradative pathway is the isomerization of the aldehyde group of the reducing moiety to a keto group. If the latter residue is substituted at carbon 4 a $\beta$-elimination reaction ensues and the reaction product from the residue undergoes rearrangement to isosaccharinic acid. A polysaccharide fragment with one less residue is formed. Depending on the reaction conditions, the isosaccharinic acid may undergo decomposition to hydroxy acetic acid, glycolic acid, and other products and eventually to $CO_2$.

Alkaline degradation of polysaccharides generally does not go to completion because in addition to elimination reactions of the above type chain termination reactions can occur if substituents on the reducing residue at carbon 3 are eliminated. The latter reactions occur when the hydroxyl group at carbon-3, due to rearrangements of the carbonyl and hydroxyl groups at carbon 1 and 2, is eliminated rather than the substituent group at carbon 4. The reaction sequence leads to the formation of the stable metasaccharinic acid group. Termination reactions can be used to locate the position of substituent groups on the reducing residue and to establish the position of branch points in polysaccharides.

It is clear that treatment of polysaccharides with alkali brings about many changes including isomerization, oxidation, reduction, and molecular rearrangements of reducing residues and the fragmentation of the polysaccharide chains. The products are numerous and varied in types and may consist of formic acid, acetic acid, glycolytic acid, lactic acid and saccharinic acids and $CO_2$. Available information on the alkaline degradation of polysaccharides has been well summarized (23) but only a limited amount of structural information can be obtained by this method because of the complex nature of the reactions.

### 7.7.2 *$\beta$-Elimination reactions*

Carbohydrate residues with substituent groups located in the $\beta$-position to electron withdrawing groups undergo $\beta$-elimination reactions on treatment with a base (23). The substituent groups which can be eliminated include alkoxyl, hydroxyl or glycosyl and the electron withdrawing groups may be aldehydic, carbonyl, carboxylic acid esters, amides or sulphones. In order for the elimination reaction to occur a hydrogen must be present in the $\alpha$-position to the electron withdrawing group. The $\beta$-elimination reactions have been especially useful in structural studies on polysaccharides, particularly on polysaccharides containing uronic acid residues. Reactions of this type are also utilized in structural studies on the carbohydrate moiety of glycoproteins with *O*-glycosidic linkages.

In order for $\beta$-elimination reactions to occur in uronic acid containing polysaccharides, the carboxyl group of the uronic acid must be esterified to form an electron withdrawing group and the residue must be substituted at the $\beta$-position to the ester grouping. The reaction sequence for the elimination is shown in Equation 5.

$$(5)$$

The initial reaction of the sequence is the formation of a double bond between carbons 4 and 5 of the uronic acid accompanied by the elimination of the substituent at carbon 4. Subsequently mild acid hydrolysis is used to release the substituent at carbon 1 and degrade the uronic acid substituent. Carbohydrate moieties which are attached to carbon 1 or 4 of the uronic acid will be eliminated. Both types of moieties can be isolated and used for further structural studies. The position of attachment of the original uronic acid units can be determined by further methylation of the carbohydrate fragments and identifying the new methyl derivative which is produced. Results of uronic acid elimination reactions and subsequent analysis for a bacterial polysaccharide (*Formula 1*) are shown in *Figure 10* establishing that the uronic acid units are attached to carbon 4 of the rhamnose residues.

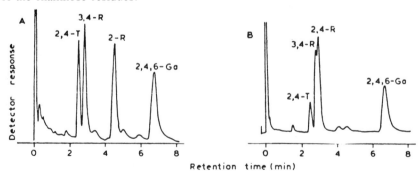

**Figure 10.** Gas liquid chromatography patterns for the native polyaccharide of *S. bovis* (**Frame A**) and the polyaccharide after $\beta$-elimination of the glucuronic acid units and remethylation (**Frame B**): 2,4-T = 1,3,5-tri-*O*-acetyl-2,4-di-*O*-methyl-deoxy talitol, 3,4-R = 1,2,5-tri-*O*-acetyl-3,4-di-*O*-methyl rhamnitol, 2-R = 1,3,4,5-tetra-*O*-acetyl-2-*O*-methyl rhamnitol, 2,4,6-Ga = 1,3,5-tri-*O*-acetyl-2,4,6-tri-*O*-methyl galactitol, and 2,4-R = 1,3,5-tri-*O*-acetyl-2,4-tri-*O*-methyl rhamnitol (24).

A polysaccharide of galactose, rhamnose and internal glucuronic acid units from *Klebsiella* was subjected to $\beta$-elimination reactions. The data which were obtained coupled with earlier data led to the elucidation of the structure as shown in *Formula 8* for the repeating unit of this polysaccharide.

$$\rightarrow 3)\text{-}D\text{-Gal}p\text{-}(1\rightarrow 2)\text{-}L\text{-Rha}p\text{-}(1\rightarrow 4)\text{-}D\text{-GlcA}p\text{-}(1\rightarrow 3)\text{-}D\text{-Gal}p\text{-}(1\rightarrow 4)\text{-}L\text{-Rha}p\text{-}(1\rightarrow$$

$$2$$
$$\uparrow$$
$$1$$
$$D\text{-Gal}p$$

Formula 8

The $\beta$-elimination reaction can be coupled with other degradative reactions and used for obtaining additional information on the structure of a polysaccharide (23,29). The oxidation of the $\beta$-elimination product from a uronic acid containing compound can be effected at the hydroxyl group converting the hydroxyl group to a carbonyl group. Ruthenium tetraoxide has been used as the oxidizing agent. The product of methylation, $\beta$-elimination and oxidation of a uronic acid containing polysaccharide (29), after treatment with base followed by sodium borohydride was subjected to further methylation, hydrolysis, reduction and acetylation. An analysis by g.l.c.-m.s. showed that 3-O-acetyl-1,2,4,5,6-penta-O-methyl galactitol, 1,5-di-O-acetyl-2,3,4-tri-O-methyl rhamnitol and 1,3,5-tri-O-acetyl-2,4-di-O-methyl-6-deoxy talitol were produced and the monosaccharide sequence of this polysaccharide as shown in *Formula 1* is thereby verified.

The experimental details for the elimination of glucuronic acid and subsequent oxidation with ruthenium tetraoxide are presented for a polysaccharide of 6-deoxy-talose, rhamnose, galactose and glucuronic acid (24,29).

(i)     Methylate the hydroxyl groups and esterify the carboxyl groups of 12 mg of streptococcal polysaccharide by the procedure described in Section 6.

(ii)    Purify the product on Sephadex LH-20 and transfer to a small flask.

(iii)   Evaporate to dryness under nitrogen in a water bath at 35°C.

(iv)    Fit the flask with a serum cap and flush with nitrogen.

(v)     Add the following reagents by syringes inserted through the serum cap: first, 2 ml of a solution (19:1 by vol.) of dry dimethylsulphoxide and 2,2-dimethoxypropane containing a trace amount of *p*-toluene sulphonic acid and second, 0.5 ml of freshly prepared methylsulphinyl methyl sodium.

(vi)    Sonicate the mixture for 30 min and maintain at room temperature overnight.

(vii)   Place the reaction mixture in an ice bath and stop the reaction by addition of excess 50% acetic acid.

(viii)  Add the complete reaction mixture to 5 ml of water and extract the methylated and modified polysaccharide three times with 10 ml portions of chloroform.

(ix)    Wash the combined chloroform extracts three times with water.

(x)     Divide the extract into two equal portions.

(xi)    Evaporate one portion to dryness and save the other for subsequent analysis.

(xii)   Subject the residue obtained to mild hydrolysis in 2 ml of 10% acetic acid in a sealed ampoule under nitrogen for 1 h at 100°C.

(xiii)  Dry the reaction mixture by lyophilization and purify the polymeric residue by gel filtration on Sephadex LH-20.

(xiv)  Evaporate the solvent and remethylate by the standard procedure.
(xv)   Hydrolyse and convert the products to the alditol acetates as described in Section 6.

The products identified in the mixture were 1,3,5-tri-*O*-acetyl-2,4-di-*O*-methyl-6-deoxy talitol, 1,2,5-tri-*O*-acetyl-3,4-di-*O*-methyl rhamnitol, 1,3,5-tri-*O*-acetyl-2,4-di-*O*-methyl rhamnitol and 1,3,5-tri-*O*-acetyl-2,4,6-tri-*O*-methyl galactitol.

(xvi)    Dissolve the other portion of the methylated and modified polysaccharide in 3 ml of methylene dichloride in a 50-ml round bottom flask and add 13 ml of a 0.1 M solution of ruthenium tetraoxide in methylene chloride from a dry glass syringe.
(xvii)   Next add 16 ml of saturated aqueous solution of sodium meta periodate.
(xviii)  Stopper the flask and shake vigorously until both solvent phases remain yellow in colour indicating that the reaction is complete.
(xix)    Remove the aqueous layer containing the periodate.
(xx)     Shake the organic phase with 0.2 ml of isopropyl alcohol to convert the ruthenium tetraoxide to ruthenium dioxide.
(xxi)    Remove the precipitate of ruthenium dioxide and other impurities by washing the organic phase several times with 8 ml portions of water.
(xxii)   Extract the combined aqueous washings with 8 ml of methylene chloride and combine the methylene chloride phase and the original organic phase.
(xxiii)  Filter through glass wool.
(xxiv)   Evaporate the solvent under nitrogen at 40°C.
(xxv)    Dissolve the material in 3 ml of methylene chloride and add 2 ml of freshly prepared 1.0 M sodium methoxide solution. Maintain the reaction mixture at room temperature for 4 h to liberate the oligosaccharide fragments from the polysaccharide.
(xxvi)   Acidify to pH 4 by adding 90% acetic acid. Remove the solvents by evaporation under nitrogen at 45°C.
(xxvii)  Hydrolyse the modified polysaccharide in 8 ml of 50% acetic acid for one h at 105°C.
(xxviii) Dilute the hydrolysate with one volume of water, freeze in liquid nitrogen and lyophilize to dryness.
(xxix)   Partition the residue between methylene chloride and water phases.
(xxx)    Wash the organic phase three times with 3 ml of water and dry over $MgSO_4$. Remove the $MgSO_4$ by centrifugation.
(xxxi)   Filter the methylene chloride layer through glass fibre paper and evaporate the solvent under nitrogen.
(xxxii)  Reduce the methylated oligosacharide by dissolving the sample in 5 ml of a mixture of dioxane and ethanol (8:3 v/v) and adding 50 mg of sodium borohydride. Maintain the reaction mixture for 24 h at room temperature in which time the pH increases from 8.0 to 10.5.
(xxxiii) Acidify by adding Dowex 50 hydrogen ion form until the pH drops to 4.0.
(xxxiv)  Remove the resin by filtration through glass fibre paper and wash the resin several times with methanol. Combine the washings and the filtrate from the glass fibre paper and remove the solvent under reduced pressure at 40°C.

(xxxv) Remove the borate by repeated evaporation of methanol from the residue.

(xxxvi) Dry the methylated and reduced oligosaccharide in a vacuum desiccator overnight and remethylate by the procedure described in Section 6.4.

The fully methylated alditol of the oligosaccharide was hydrolysed, reduced and acetylated. The products were analysed by g.l.c. and by m.s. and found to be 3-O-acetyl-1,2,4,5,6-penta-O-methyl galactitol, 1,5-di-O-acetyl-2,3,4-tri-O-methyl rhamnitol and 1,3,5-tri-O-acetyl-2,4-di-O-methyl-6-deoxy talitol (29). The structure for the repeating unit of the polysaccharide is therefore as shown in *Figure 1*.

### 7.7.3 Degradation by deamination

The most common 2-acetamido-2-deoxy hexoses found in biological materials are 2-acetamido-2-deoxy-D-glucose (*N*-acetyl glucosamine) and 2-acetamido-2-deoxy-D-galactose (*N*-acetyl galacrosamine). Less common is 2-acetamido-2-deoxy-D-mannose (*N*-acetyl mannosamine) which occurs as a constituent unit of some polysaccharides and as a structural unit of sialic acids that are constituents of glycoproteins and have important biological roles in glycoprotein transport. Generally the *N*-acetyl amino sugars occur as constituents of heteropolysaccharides but chitin, a polymer of only *N*-acetyl glucosamine, is a notable exception. For procedures used in structural investigations of amino-sugar containing polysaccharides the amino sugar is deacetylated in the initial step of the analysis. The deacetylation is performed under mild basic conditions to minimize removal of other labile groups and prevent extensive degradation reactions (30). Barium hydroxide in water, sodium ethoxide in ethanol or dry hydrazine have been used in the past and the use of trifluoroacetic acid and trifluoroacetic anhydride has been introduced recently. Some decomposition of the amino sugars can occur with some of these reagents as for example with hydrazine. Hydrolysis in hydrazine sulphate minimizes these decomposition reactions. The next reaction for characterizing the amino sugar polysaccharides involves the deamination of the amino sugar with nitrous acid. In this reaction, a product with a 2,5 anhydro ring and a free aldehyde group at carbon 1 is formed. The reaction sequence is illustrated in Equation 6.

In the reaction the amino sugar molecule undergoes a rearrangement because of the attachment of the intermediate diazonium ion on the pyranose ring oxygen. Very often an inversion of configuration occurs at carbon 2. If the amino sugars are constituents of polysaccharides, the glycosidic bonds are also cleaved simultaneously thereby yielding polysaccharide fragments useful in further structural studies.

Applications of the deamination reaction in structural analysis of polysaccharides are varied. Chitin was deacetylated with hydrazine and deaminated with nitrous acid and the major product was 2,5-anhydromannose. Heparin, a polysaccharide composed of *N*-acetyl-glucosamine and glucuronic acid has been investigated. A number of details on the fine structure of this polysaccharide were clarified by these studies. The amino sugar degradation reactions have also been useful for investigating the structure of the carbohydrate moieties of glycoproteins with specific biological activities.

## 8. DETERMINATION OF THE RING STRUCTURE OF COMPONENT MONO-SACCHARIDES

In solution monosacharides occur in the open chain, the pyranose and the furanose ring structures. These structural forms are in equilibrium and the concentration of an indivdual form is characteristic of the monosaccharide type. When monosacharides are joined together by glycosidic linkages to form polysaccharides, the ring structures become fixed for all residues except the reducing end residue. In nature polysaccharides may contain a single type of ring structure or a mixture of ring types. With the latter compounds ring structural types are more difficult to deduce.

In the elucidation of the complete structure of a polysaccharide it is necessary to establish the type of ring structure of each monosaccharide unit. Residues which possess a furanose structure and which are terminal units are removed by mild acid hydrolysis. Following hydrolysis the monosaccharide which may be released is isolated and identified. If residues with furanose structure are internal residues then a fragmentation of the polysaccharides will occur at the locations of such residues. In this reaction the furanose residues will become reducing units of the fragments. Such polysaccharide fragments can be used in additional structural studies.

Monosaccharides with different ring structures will yield characteristic products on methylation analysis. Thus on methylation analysis the derivatives from residues with furanose rings will contain methoxyl groups at carbon 5 while residues with a pyranose ring structure will yield methoxyl groups at carbon 4. The methylation technique and the identification of the methyl derivatives can be performed as described in Section 5.1. In such an analysis the monosaccharide units with pyranose rings which are substituted at carbon 4 would yield the same derivative as residues with a furanose ring but substituted at carbon 5. To establish the structure of a residue with a pyranose ring substituted at carbon-4 and a furanose ring substituted at carbon-5 additional alkylation analysis involving different alkylation reagents can be performed. Such analysis utilizes both methylation and ethylation reactions (27).

The polysaccharide is first fully methylated and subjected to partial acid hydrolysis. Hydrolysis conditions are selected to yield a mixture of methylated oligosaccharides. The methylated oligosaccharides are subjected to ethylation with ethyl iodide and methylsulphinyl methyl sodium. The derivatized products are resolved into individual oligosaccharide derivatives by h.p.l.c. Hydrolysis, reduction and acetylation of the oligosaccharide derivatives is effected. The products are separated and identified by g.l.c. and m.s. In an analysis of the above type a 4-substituted glucose residue would yield 1,5-di-*O*-acetyl-4-*O*-ethyl 2,3,6-trimethyl glucitol while a 5-substituted glucose residue would yield 1,4-di-*O*-acetyl-5-*O*-ethyl, 2,3,6-tri-*O*-methyl glucitol. The residues with the furanose and the pyranose rings are thereby identified. It is possible to differentiate between 4-substituted pyranose units and 5-substituted furanose units by a direct analysis of the alkylated oligosaccharides in the mass spectrometer using chemical ionization methods.

## 9. DETERMINATION OF THE ANOMERIC CONFIGURATION OF GLYCOSIDIC LINKAGES

### 9.1 Enzymic hydrolysis

Susceptibilty of polysaccharides to hydrolysis by enzymes of known specificity is an excellent method for determining the configuration of the glycosidic linkages but enzymes of known specificity or purity are not always available. Some polysaccharides contain glycosidic linkages with only one type of anomeric configuration but others contain both types of anomeric linkages. Consequently more than one type of enzyme will need to be employed in determination of the configuration of the linkages in the latter compounds. It should be stressed that to utilize enzymic procedures enzymes with specificity for a single type of anomer are needed as described in Section 7.6. It is ideal to have enzymes which sequentially and selectively remove residues from a polysaccharide and would also establish the configuration of linkages. In such procedures methods for determining the types and the number of residues that are removed are also needed.

### 9.2 Nuclear magnetic resonance spectroscopy

The $^1$H-n.m.r. and $^{13}$C-n.m.r. spectra of polysaccharides have been recorded and the data used to identify anomeric configuration of the glycosidic linkages of polysaccharides (27,31,32). For such measurements native polysaccharides and derivatives of the constituent units are needed. Chemical shifts in the spectra are attributable to the orientation of groupings in the molecule. Different shifts in the signals can be obtained for $\alpha$- and $\beta$-anomeric linkages and assignment of configuration of linkages in an unknown polysaccharide can be made. The assignment of configuration unambiguously is possible only if the spectra of suitable reference compounds are available. Also if a single type of linkage is present in the polysaccharide then the assignments are more apparent.

$^{13}$C-n.m.r. provides information on anomeric configuration but also on other aspects of polysaccharide structure such as monosaccharide composition, the monosaccharide sequence and the conformation of the polysaccharide. The use of the Fourier transform methodology has greatly increased the sensitivity of this method and spectra of polysaccharides can be obtained with the natural abundance of $^{13}$C. In $^{13}$C-n.m.r. measurements it is necessary to make measurements on standard compounds and to make comparison of chemical shifts for corresponding carbon atoms. Such measurements and comparisons have shown that the chemical shifts in monosaccharides are essentially the same as the shifts obtained with a polysaccharide of the same units (31). Also it has been found that substituent groups attached to a carbon atom will cause a large increase in the chemical shift of this atom and such differences indicate the location of the glycosidic linkages.

Polysaccharides which possess a linear structure and are free of substituent groups yield relatively simple spectra from which structural features can be deduced. The $^{13}$C-n.m.r. spectra of a polysaccharide of *N*-acetyl mannosamine shown in *Figure 11* illustrates some of these features. Measurements of the spectra of this polysaccharide and for those of model compounds have been made. Calculations on chemical shifts

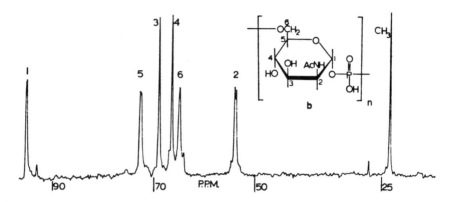

**Figure 11.** $^{13}$C-n.m.r. resonance spectrum for the *O*-deacetylated polysaccharide of *N*-acetyl mannosamine from *Neisseria meningitidis* as measured in a Varian XL-100 Fourier transform spectrometer (31).

**Table 5.** $^{13}$C-N.m.r. chemical shifts of the *O*-deacetylated polysaccharide (Polysac) of *N*-acetyl mannosamine and the glycosides of α- and β-*N*-acetyl mannosamine (α-anomer and β-anomer) (31).

| Compound | C-1 | C-2 | C-3 | C-4 | C-5 | C-6 |
|---|---|---|---|---|---|---|
| Polysac | 96.4 | 54.3 | 69.7 | 67.1 | 73.5 | 65.6 |
| α-Anomer | 94.3 | 54.4 | 70.1 | 68.0 | 73.2 | 61.7 |
| β-Anomer | 94.3 | 55.3 | 73.2 | 67.8 | 77.5 | 61.7 |

were made and these are recorded in *Table 5*. Comparisons of the differences in the chemical shifts for the *O*-deacetylated polysaccharide and for the derivatives of the α- and β-*N*-acetyl mannosamine show that there is a very large difference in the shift at carbon-6 for the polysaccharide and the two derivatives. These differences establish that the polysaccharide possesses $(1 \rightarrow 6)$ linkages. The data in *Table 5* show that only small differences in chemical shifts occurred at carbons 2, 3 and 5 for the polysaccharide and the α-anomer but in contrast large differences in shifts for the β-anomer and the polysaccharide were observed. These results establish that configuration of the glycosidic linkages in the polysaccharide is α.

## 9.3 Chromium trioxide oxidation

The configuration of the glycosidic linkages of polysaccharides can be determined by chromium trioxide oxidation followed by methylation analysis of the oxidized polysaccharide (23). It has been demonstrated with model compounds that oxidation with chromium trioxide of anomers will differentiate the α and β configuration. Thus glycosides in which the aglycone moieties occupy the equatorial (the β-glycosidic) position are oxidized much faster than glycosides in which the aglycone occupies the axial (α-glycosidic) position. The difference has been attributed to the ease of formation of a ketoester by cleavage at the bridge oxygen of β-anomeric compounds. A polysaccharide with α-glycosidic linkages is not or only slowly oxidized and hence on methylation analysis yields essentially the same types of derivatives as those from the native polysaccharide. A polysaccharide with β-glycosidic linkages is oxidized rapidly and carbohydrate residues are degraded. As a result g.l.c. patterns for the methylated fragments are not obtained or are markedly modified.

The chromium trioxide oxidation method has been used for determining the configuration of the glycosidic linkages in a polysaccharide of rhamnose and glucose from *S. bovis*.

(i)     Dissolve 5 mg of the polysaccharide in 1 ml of $N,N^1$-dimethylformamide and mix with 1 ml of an equal part mixture of pyridine and acetic anhydride.

(ii)    Acetylate the polysaccharide at room temperature overnight.

(iii)   Recover the products by chromatography on a column of Sephadex LH-20, with acetone as the eluting solvent.

(iv)    Identify the fractions containing the polysaccharide acetate by a suitable colorimetric test.

(v)     Combine and evaporate these fractions under a stream of nitrogen.

(vi)    Subject the derivative to oxidation by adding 25 mg of finely powdered chromium trioxide in 0.2 m of glacial acetic acid. Maintain the reaction mixture at 50°C for 1 h.

(vii)   Re-isolate the polysaccharide by chromatography on Sephadex LH-20 as described above.

(viii)  Methylate the chromium trioxide-treated polysaccharide as described in Section 6.4.

(ix)    Hydrolyse, reduce and acetylate the product and analyse by g.l.c.

The pattern obtained for the chromium trioxide treated and methylated polysaccharide was identical to the pattern obtained for the native polysaccharide and therefore the configuration of the glycosidic linkage is established to be $\alpha$.

## 10. IMMUNOLOGICAL METHODS

That polysaccharides can stimulate the immune system to produce antibodies was discovered many years ago. The early work was concerned with the antigenicity of polysaccharides from the cell wall of microorganisms (33) and the development of serological methods for identifying certain groups of bacteria (34). It is now well known that polysaccharides and glycoproteins, in general, induce an immune response in which antibodies specific for the carbohydrate residues of the antigen are synthesized (35,36). Antibodies are proteins of unique structure and with special arrangements of the polypeptide chains. Recent work has shown that the antibody molecule consists of two light and two heavy polypeptide chains arranged in a structure in which segments of a light and heavy chain form a combining site. The antigen combines with the antibody to form a precipitin complex which can be observed by capillary precipitin, agar diffusions and enzyme linked immunoassays. A laboratory protocol for performing agar diffusion is given below.

(i)     Layer 3 ml of 1% agarose solution in a microscope slide. Allow the agarose to gel.

(ii)    Cut holes in the gel either 2 $\mu$l or 10 $\mu$l in size with a standard punch.

(iii)   Introduce 2 $\mu$l or 10 $\mu$l of the antigen or antibodies into individual wells.

(iv)    Place the slide in a Petri dish in a moist atmosphere.

(v)     Observe the gel for formation of precipitin bands for the next 36 h. Photograph the slide showing bands for a permanent record.

*Figure 12* shows a photograph of a slide and the appearance of the preciptin bands

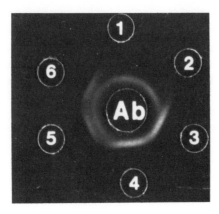

**Figure 12.** A double diffusion agar gel pattern for antigenic polysaccharides from *S. faecalis*: **wells 1** and **2** polysaccharide I, **wells 3** and **4** polysaccharide II, and **wells 5** and **6** mixture of the two antigenic polysaccharides.

for antigenic polysaccharides in an extract from the cell wall of *S. faecalis* and the homologous antiserum.

Immunological methods can be used in several ways for the structural characterization of polysaccharides. Vaccines used to immunize animals generally contain a mixture of antigens and accordingly a heterogeneous population of antibodies will be obtained. These antibodies in the serum can be used to determine the number of polysaccharides and the purity of these polysaccharides. In the photograph in *Figure 12*, well Ab contains the antibodies, wells 5 and 6 contain an extract of carbohydrate antigens from the cell wall of a Streptococcus. The slide shows that two bands of precipitin complex are formed and therefore the extract contains two different antigenic polysaccharides. It is possible to separate the two antigenic polysaccharides by Bio-Gel chromatography. Wells 1 and 2 contain one polysaccharide and wells 3 and 4 contain the other. It can be seen that single bands of precipitin complex are obtained with each antigenic polysaccharide. These precipitin bands are at different positions showing that the antigens are different compounds. Further, the single band with each antigen indicates that each polysaccharide has been purified to a high degree by the chromatography procedure. The fusion of the bands between wells 1 and 2 and between wells 3 and 4 shows identity of the antigens in each pair of wells. The spurs between wells 1 and 6 and between wells 2 and 3 indicate differences in structure of the two polysaccharides.

Many types of anti-carbohydrate antibodies have been isolated in pure form by affinity chromatography methods (37) and the specificity of some of these antibodies for carbohydrate units has been established by periodate oxidation and hapten inhibition techniques (36,38). Oxidation of the antigen with periodate destroys the carbohydrate units and also destroys the antigenicity. A hapten group of an antigen is that grouping of functional groups which combine with the antibody combining site. The nature of this group can be determined by inhibition methods. Polysaccharides with similar determinant groups may precipitate with antibodies specific for other antigens. This type of precipitin formation has been called cross reactivity and has been used to identify similar structural features of polysaccharides. The extent of cross reactivity indicates

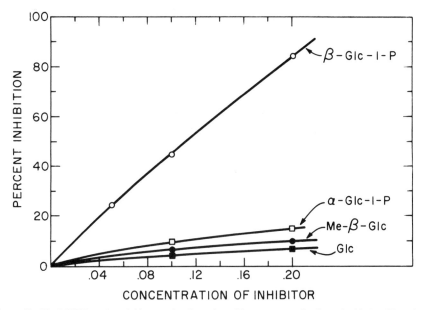

**Figure 13.** The inhibition of precipitin complex formation of the streptococcal polysaccharide I and homologous antiserum by various carbohydrates (38).

the degree of similarity in structure of the polysaccharides (39).

In recent years the immunological technique has been extremely valuable and popular for detecting and identifying cellular constituents *in situ*. Antibodies are quite specific for a single type of structural group and as a result can be used for identification purposes. The specificity of antibody reactions can be further enhanced by use of monoclonal antibodies in which a single type of molecular species exists. The monoclonal antibodies combine with antigens in a similar way as polyclonal antibodies.

Hapten inhibition techniques have also been used for identification of structural features of a polysaccharide (38). Some results of hapten inhibition experiments with antibodies specific for $\beta$-glucose-1-phosphate are shown in *Figure 13*. It is noted in the Figure that $\beta$-glucose-1-phosphate inhibits over 80% of the precipitin reaction at the concentrations tested. On the other hand, glucose, methyl glucoside and $\alpha$-glucose-1-phosphate are not effective as inhibitors causing less than 10% inhibition. Therefore, the determinant groups of this antigen are $\beta$-glucose-1-phosphate units. This type of inhibiton experiment is of general applicability to all other antigen-antibody systems.

Recently it has been possible to purify antibodies by affinity chromatography and such antibodies can be used as diagnostic agents for some types of diseases and as therapeutic agents for others. To purify antibodies by affinity chromatography, it is necessary to determine first the nature of the immunodeterminant group of the antigen. Second it is necessary to couple this group to an inert support and prepare an affinity adsorbent.

The preparation of vaccines, the immunization regimes, the synthesis of the immuno-adsorbent with carbohydrate ligands and the techniques of affinity chromatography are described for two antigens, a bacterial polysaccharide of glucose and galactose and a glycoconjugate of galactose and bovine serum albumin.

(i)     Grow *S. faecalis* strain N in 10 ml of a 3% solution of Todd-Hewitt broth at 37°C for 24 h.

(ii)    On successive days inoculate 75 ml of Todd-Hewitt broth with the 10-ml culture and then 500 ml of broth with the 75-ml culture. Grow all cultures at 37°C for 24 h.

(iii)   Collect the cells from the 500-ml culture by centrifugation at 4°C for 15 min.

(iv)    Wash the pellet three times with 100 ml of 0.2% solution of formaldehyde in saline (0.9% NaCl). Recover the cells after each washing by centrifugation.

(v)     Stir the cells in 100 ml of 0.2% formaldehyde in saline solution at 4°C for 24 h.

(vi)    Recover the cells by centrifugation and suspend the cells in 80 ml of sterile saline solution.

(vii)   Perform viability tests with the formaldehyde treated cells to check that the cells are non-viable.

(viii)  Measure the absorption at 600 nm. Absorption values from 1.0 to 1.5 should be obtained.

(ix)    Store the vaccine at 4°C.

   Glycoconjugates of carbohydrates and proteins also induce anti-carbohydrate antibody synthesis in rabbits.

(i)     Dissolve 15 mg of galactosyl-bovine serum albumin in 5 ml of buffered saline and mix with 5 ml of Freunds complete adjuvant.

(ii)    Immunize rabbits with the bacterial vaccine by injecting 0.2 ml of the vaccine intravenously into the ear vein for three consecutive days. Follow with a rest period of 4 days.

(iii)   Repeat the routine for periods of 15 to 20 weeks with a rest period of one month following the 6th week.

(iv)    Obtain blood samples from the ear vein at the end of each rest period commencing on the 7th week of immunization.

(v)     Immunize rabbits with the glycoconjugate vaccine by subcutaneous injection of 0.2 ml of the vaccine in the back of the neck. Immunize for 3 weeks and follow with a rest period of 3 weeks.

(vi)    Secondary immunization is performed with the glycoconjugate by repeating the above sequence.

(vii)   Bleed rabbits during the secondary immunization. Allow the blood to clot for 3 h at room temperature and for an additional 24 h at 4°C.

(viii)  Centrifuge the sample and collect the serum carefully with a pipette. Store the serum in sterile polyethylene tubes at $-20$°C.

   To prepare a carbohydrate affinity adsorbent (35) carry out the following protocol.

(i)     Swell 12 g of cyanogen bromide-actived Sepharose 4B in dilute acid by washing the Sepahrose on a glass sintered filter with 700 ml of 0.001 M HCl.

(ii)    Next wash the activated Sepharose with 200 ml of 0.1 M NaHCO$_3$ adjusted to pH 9.0 with NaOH.

(iii)   Dissolve 170 $\mu$mol of *p*-aminophenyl $\beta$-lactoside or 85 $\mu$mol of *p*-aminophenyl $\beta$-galactoside in 20 ml of 0.1 M NaHCO$_3$ (pH 9.0) and mix each solution with 12 g of the swollen and moist Sepharose.

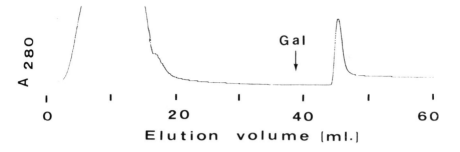

**Figure 14.** Affinity chromatography pattern of antibodies directed at galactose bovine serum albumin. The arrow indicates the point of introduction of 0.5 M galactose solution (36).

(iv)   Shake the mixture for 24 h on a wrist action shaker at 4°C.

(v)    Recover the reaction product on a glass sintered filter and wash with 150 ml of 0.1 M NaHCO₃ (pH 9.0).

(vi)   Shake the reaction product with 100 ml of 1 M ethanolamine (pH 9.0) for 2 h at room temperature.

(vii)  Wash with 0.1 M phosphate buffer (pH 7.2) in saline. The former adsorbent will contain *p*-aminophenyl-lactosyl units and the latter *p*-aminophenyl galactosyl units linked via the amino groups to the imidocarbonate groups of the activated Sepharose.

(viii) Pour the glycosyl-Sepharose in a 30 × 2.5 cm column and equilibrate with 0.1 M phosphate buffer (pH 7.2) in saline at room temperature.

Use the columns to isolate anti-carbohydrate antibodies with specificity for lactose and for galactose from appropriate antiserum.

The protocol for isolation of anti-galactose antibodies from the serum of a rabbit immunized with galactosyl bovine serum albumin (36) is as follows.

(i)    Introduce a sample of 2 ml of antiserum on a column of galactosyl Sepharose washed well with buffer solution.

(ii)   Wash the column with 0.1 M phosphate buffer (pH 7.2) in saline and monitor the eluate with an u.v. analyser and flow cell.

(iii)  When the serum protein has passed through the column, elute the adsorbed antibodies with 10 ml of 0.5 M galactose solution. The u.v. absorption curve for the eluates is shown in *Figure 14*. The arrow indicates the point of introduction of the galactose solution. The second peak is due to the elution of the anti-galactose antibodies.

(iv)   Collect these eluates and precipitate the antibodies with an equal volume of saturated ammonium sulphate. Recover the antibodies by centrifugation.

(v)    Dissolve the antibodies in 1.0 ml of 0.1 M phosphate buffer (pH 7.2) in saline.

(vi)   Combine such anti-galactose antibody preparations from several affinity columns.

(vii)  Dialyse against 0.1 M phosphate buffer (pH 7.2) in saline for 72 h at 4°C.

(viii) Concentrate by lyophilization and store the antibodies at −20°C until used for other experiments.

## 12. ACKNOWLEDGEMENTS

The author acknowledges the many contributions of his former and present students, particularly L.S.Forsberg, K.L. Dreher, B.M. Romanic, Y.Tominaga, S.A.Kelly-Delcourt and F.J.Miskiel, to the methods described in this chapter.

## 13. REFERENCES

1. Pigman,W. and Horton,D. (1970) *The Carbohydrates*, 2nd Edition, Academic Press Inc., London and New York, Vol. **II A**, p. 375 and Vol. **II B**, p. 471.
2. Joint Commission on Biochemical Nomenclature (1982) *J. Biol. Chem.*, **257**, 3352.
3. Block,R.J., Darrum,E.L. and Zneig,G. (1958) *Paper Chromatography and Paper Electrophoresis*, 2nd Edition, Academic Press Inc., New York, p. 170.
4. French,D., Knapp,D.W. and Pazur,J.H. (1950) *J. Am. Chem. Soc.*, **72**, 5150.
5. Kennedy,J.F. and Fox,J.E. (1980) in *Methods in Carbohydrate Chemistry*, Whistler,R.L. and BeMiller,J.N. (eds.), Academic Press Inc., New York and London, Vol. **8**, p. 3.
6. DuBois,M., Gilles,K.A., Hamilton,J.K., Rebers,P.A. and Smith,F. (1956) *Anal. Chem.*, **28**, 350.
7. Pazur,J.H., Dropkin,D.J., Dreher,K.L., Fosberg,L.S. and Lowman,C.S. (1976) *Arch. Biochem. Biophys.*, **176**, 257.
8. Pazur,J.H. and Kleppe,K. (1964) *Biochemistry*, **3**, 578.
9. Whistler,R.L. and Anisuzzaman,K.M. (1980) in *Methods in Carbohydrate Chemistry*, Whistler,R.L. and BeMiller,J.N. (eds.), Academic Press, New York and London, Vol. **8**, p. 45.
10. Schachman,H.K. (1957) in *Methods in Enzymology*, Colowick,S.P. and Kaplan,N.O. (eds.), Academic Press Inc., New York, Vol. **4**, p. 32.
11. Pazur,J.H., Kleppe,K. and Anderson,J.S. (1962) *Biochim. Biophys. Acta*, **65**, 369.
12. Martin,R.G. and Ames,B.N. (1961) *J. Biol. Chem.*, **236**, 1372.
13. Isbell,H.S. (1973) in *Methods in Carbohydrate Chemistry*, Whistler,R.L. (ed.), Academic Press Inc., New York and London, Vol. **5**, p. 249.
14. Haworth,W.N. (1915) *J. Chem. Soc. (Lond.)*, **107**, 8.
15. Hakomori,S. (1964) *J. Biochem. (Tokyo)*, **55**, 205.
16. Bjorndal,H., Hellerqvist,C.G., Lindberg,B. and Svensson,S. (1970) *Angew. Chem. Internat. Edit.*, **9**, 610.
17. Lindberg,B. (1972) in *Methods in Enzymology*, Ginsburg,V. (ed.), Academic Press Inc., New York and London, Vol. **28**, p. 178.
18. Dutton,G.G.S. (1973) *Adv. Carbohydr. Chem. Biochem.*, *28*, 12.
19. Pazur,J.H., Cepure,A., Kane,J.A. and Hellerqvist,C.G. (1973) *J. Biol. Chem.*, **248**, 279.
20. Seymour,F.R., Plattner,R.D. and Slodki,M.E. (1975) *Carbohydr. Res.*, **44**, 181.
21. Hay,G.W., Lewis,B.A. and Smith,F. (1973) in *Methods in Carbohydrate Chemistry*, Whistler,R.L. (ed.), Academic Press Inc., New York and London, Vol. **5**, p. 347 and p. 361.
22. Pazur,J.H. and Forsberg,L.S. (1977) *Carbohydr. Res.*, **58**, 222.
23. Lindberg,B., Lonngren,J. and Svensson,S. (1975) *Adv. Carbohydr. Chem. Biochem.*, **31**, 185.
24. Pazur,J.H. and Forsberg,L.S. (1978) *Carbohydr. Res.*, **60**, 167.
25. Stewart,T.S. and Ballou,C.E. (1968) *Biochemistry*, **7**, 1855.
26. Wong,C.G., Sung,S.S.J. and Sweeley,C.C. (1980) in *Methods in Carbohydrate Chemistry*, Whistler,R.L. and BeMiller,J.N. (eds.), Academic Press Inc., New York and London, Vol. **8**, p. 55.
27. McNeil,M., Darvill,A.G., Aman,P., Franzen,L.-E. and Albersheim,P. (1982) in *Methods in Enzymology*, Ginsberg,V. (ed.), Academic Press Inc., New York and London, Vol. **83**, p. 3.
28. Yamashita,K., Mizuochi,T. and Kobata,A. (1982) in *Methods in Enzymology*, Ginsberg,V. (ed.), Academic Press Inc., New York and London, Vol. **83**, p. 105.
29. Pazur,J.H. and Forsberg,L.S. (1980) *Carbohydr. Res.*, **83**, 406.
30. Kenne,L. and Lindberg,B. (1980) Kenne,L. and Lindberg,B. (1980) *Methods in Carbohydrate Chemistry*, Whistler,R.L. and BeMiller,J.N. (eds.), Academic Press Inc., New York and London, Vol. **8**, p. 295.
31. Jennings,H.J. and Smith,I.C.P. (1980) in *Methods in Carbohydrate Chemistry*, Whistler,R.L. and BeMiller,J.N. (eds.), Academic Press Inc., New York and London, Vol. **8**, p. 97.
32. Gorin,P.A.J. (1981) *Adv. Carbohydr. Chem. Biochem.*, **38**, 13.
33. Heidelberger,M. and Avery,O.T. (1924) *J. Exp. Med.*, **40**, 301.
34. Lancefield,R. (1940–1941) *Harvey Lectures Ser.*, **36**, 251.
35. Pazur,J.H., Dreher,K.L. and Forsberg,L.S. (1978) *J. Biol. Chem.*, **253**, 1832.
36. Pazur,J.H. (1982) *Carbohydr. Res.*, **107**, 243.
37. Pazur,J.H. (1981) *Adv. Carbohydr. Chem. Biochem.*, **39**, 405.
38. Pazur,J.H. (1982) *J. Biol. Chem.*, **257**, 589.
39. Heidelberger,M. (1984) *Mol. Immunol.*, **21**, 1011.

CHAPTER 4

# Proteoglycans

## S.L. CARNEY

## 1. INTRODUCTION

Proteoglycans are macromolecules which consist of a central core protein to which are attached a number of highly charged glycosaminoglycan chains. These molecules are distributed ubiquitously throughout the extracellular matrices of connective tissues. As a result of the high charge density conferred on proteoglycans by virtue of the poly-anionic glycosaminoglycan component, such molecules occupy a large volume in solution. In cartilage (and a number of other connective tissues), proteoglycans can interact specifically with hyaluronic acid forming huge multimolecular aggregates which substantially increase the hydrated volume of the proteoglycan and facilitates entrapment within the collagen network of the extracellular matrix. Many of the physiological roles of proteoglycans rely upon such space-filling properties. Hence, proteoglycans exhibit one of the most exquisite structure-function relationships in modern biology due to the complex and elegant spatial and chemical arrangement of molecular components in such biopolymers.

In the sense that proteoglycans are conjugates of protein and carbohydrate they have features in common with glycoproteins. It is the nature of the carbohydrate (glycosaminoglycan) moiety of proteoglycans which allows them to be differentiated from glycoprotein. The glycosaminoglycans of proteoglycans are high molecular weight, essentially unbranched heteropolymeric molecules, consisting of repeating disaccharides which are highly substituted with carboxyl and/or sulphate ester groups. Contrast this with the relatively low molecular weight, frequently highly branched, predominantly un-charged oligosaccharides found on glycoproteins. More recently, however, it has been demonstrated that in addition to glycosaminoglycans, proteoglycans also contain both *N*- and *O*-linked oligosaccharides similar in structure to those found on glycoproteins.

As yet, no nomenclature system has been universally accepted for proteoglycans, however, it seems probable that a system proposed by Heinegård *et al.* (1) may well be accepted in future. Such a system classifies proteoglycans according to the tissue from which they were derived, their hydrodynamic size, predominant glycosaminoglycan substituents and their ability to interact with hyaluronic acid. *Table 1* lists the proposed structure of the proteoglycans found in hyaline cartilage and characterizes them in accordance with the nomenclature of Heinegård.

Due to the wide diversity in structure of the various proteoglycans from connective tissues it may be apparent that it is not possible to outline a single set of procedures which may be used to isolate and purify all types of proteoglycan. It is possible, however, to outline a general set of procedures which may be adapted for the isolation and purifi-

**Table 1.** The proteoglycan types found in hyaline cartilage

| Proteoglycan type | Molecular weight (kd) | Proposed nomenclature | Ability to interact with HA | Proposed structure |
|---|---|---|---|---|
| Aggregating CS-rich | $3.5 \times 10^3$ | PG-LAI-CS (cart) | Yes | 97 CS, 30 KS chains protein core ~200 kd |
| Aggregating KS-rich | $1.3 \times 10^3$ | PG-LAII-CS (cart) | Yes | 41 CS, 16 KS chains protein core ~94.4 kd |
| High molecular weight; non-aggregating | $<1.3 \times 10^3$ | PG-LN-CS (cart) | No | ND |
| Low molecular weight; non-aggregating | 76 | PG.SmI-CS (cart) | No | ~1−2 CS chains protein core 42.3 kd |
| Dermatan sulphate (self-associating) | $2 \times 10^2$ | PG-SmI-DS (cart) | No | ~1 DS ~1−2 CS chains protein core 45 kd (dimer of PG.SmII-DS) |
| Dermatan sulphate (non-self-associating) | 80 − 120 | PG-SmII-DS (cart) | No | ~1 DS ~1−2 CS chains protein core 45 kd |

Abbreviations: CS, chondroitin sulphate; KS, keratan sulphate; DS, dermatan sulphate; L, high molecular weight; Sm, low molecular weight; ND, not determined.

cation of particular proteoglycans. Such general procedures for purification rely on physical differences between proteins and proteoglycans, in particular the high hydrated volume of proteoglycans which confers upon these molecules a considerably higher buoyant density in CsCl than proteins allowing them to be separated by isopycnic density gradient centrifugation in CsCl. Furthermore, the high charge density of proteoglycans allows them to be separated from protein by ion-exchange chromatographic techniques. The large hydrodynamic size may also be used to effect separation from proteins by gel permeation chromatography.

This chapter will deal with the extraction, purification and analysis of cartilage proteo-glycans, whose structure is comparatively well understood. The techniques outlined have been developed for these molecules, although one may use such procedures for purification and analysis of proteoglycans from other tissues by only minor modifi-cations to such methods.

## 1.1 The glycosaminoglycans of connective tissue

There are five main classes of glycosaminoglycans found in connective tissue: the chon-droitin sulphates, the keratan sulphates, dermatan sulphates, heparan sulphates (and heparin) and hyaluronic acid. With the exception of hyaluronic acid, all contain sulphate ester groups. Furthermore, hyaluronic acid is the only glycosaminoglycan that is not found conjugated to protein. Therefore, although hyaluronic acid may not itself be con-sidered a proteoglycan, it performs a specific function in promoting proteoglycan aggre-gation in many connective tissues.

### 1.1.1 Chondroitin sulphate

Chondroitin sulphate is found widely distributed throughout various connective tissues. It is an unbranched heteropolymer consisting of repeating disaccharides of D-glucuronic acid, linked $\beta(1 \rightarrow 3)$ to 2-acetamido-2-deoxy-D-galactose (*N*-acetyl galactosamine). The galactosamine residue may be *O*-sulphated at either C-4 (chondroitin 4-sulphate) or C-6 (chondroitin 6-sulphate). The presence of D-glucuronic acid in its structure allows chondroitin sulphate to be estimated by the carbazole assay for uronic acids (see Section 4.1). The disaccharides are linked *N*-acetyl galactosamine-$\beta$(1-4)-[glucuronic acid-$\beta$(1,3)-*N*-acetyl galactosamine]$_n$-$\beta$(1,4)-glucuronic acid. The molecular weight distribution of chondroitin sulphate depends largely on tissue source and species but generally ranges between 2.4 and 50 kd with a mean at around 20 kd. It has been demonstrated that chondroitin sulphate can possess both 4- and 6-sulphate ester groups on galactosamine residues of the same chain, thus further complicating the structure of this class of glycosaminoglycans.

## 1.1.2 *Keratan sulphate*

Keratan sulphate is found in cartilage, intervertebral disc and also the cornea. The corneal form of keratan sulphate (KSI) has the same repeating disaccharide structure as skeletal keratan sulphate (KSII) but differs in that it participates in alkali-stable linkages to protein, whereas the skeletal form has an alkali-labile linkage to threonine in the protein core of the proteoglycan. The repeating disaccharide of keratan sulphate is galactose, linked $\beta$(1-4) to *N*-acetylglucosamine. The glucosamine residue is sulphated at C-6. The disaccharides are linked *N*-acetyl glucosamine-$\beta$(1,3)-[galactose-$\beta$(1,4)-*N*-acetyl glucosamine]$_n$-$\beta$(1,3)-galactose. The absence of uronic acid from this structure prevents its estimation or detection by the carbazole reaction for uronic acids. It can, however, be detected by GAG dye binding assays (see Section 4.2) or by assays for neutral sugars (anthrone reaction) or hexosamines (Elson-Morgan assay). These reactions, however, are not specific for keratan sulphate, but may be used, for example, to monitor column effluents in conjunction with the carbazole assay for uronic acid. The presence of keratan sulphate will produce an increase in the proportion of neutral sugar (or hexosamine) compared with uronic acid.

## 1.1.3 *Dermatan sulphate*

Dermatan sulphate is an epimer of chondroitin sulphate, that is, the uronic acid residue in the repeat disaccharide is found as iduronic acid. Thus, the repeat disaccharide consists of L-iduronic acid linked $\alpha$(1-3) to *N*-acetylgalactosamine. The galactosamine residue is sulphated at C-4. Since iduronate is found in the L rather than the D configuration, the bond between it and the sugar amine is designated $\alpha$ rather than $\beta$ even though the arrangement around the C-1 atom in the uronic acid residue and the C-3 in the sugar amine is identical irrespective of whether the uronic acid is found as glucuronate or iduronate. The disaccharides are linked *N*-acetyl galactosamine-$\beta$(1,4)-[iduronate-$\alpha$(1,3)-*N*-acetyl galactosamine]$_n$-$\beta$(1,4)-iduronate. The molecular weight range for dermatan sulphate is similar to that for chondroitin sulphate although generally dermatan sulphate chains are longer than chondroitin sulphate. As with chondroitin sulphate, dermatan sulphate can be estimated by the uronic acid assay, the epimer of glucuronic acid in dermatan sulphate retains reactivity in this procedure; however, free iduronic acid or dermatan sulphate may produce as little as 25% of the colour development as that found for free glucuronic acid or chondroitin sulphate.

### 1.1.4 *Heparin*

Heparin has a more complex structure than the glycosaminoglycans described previously. The glycosaminoglycan contains both iduronic and glucuronic acid residues, and in addition many of the glucosamine residues can be N-sulphated as opposed to *N*-acetylated. In comparison with other glycosaminoglycans, heparin is particularly highly sulphated, containing up to three sulphate groups per disaccharide. There is, however, considerable variability in the sulphation of disaccharides in a single heparin chain (from 1.6 to 3 sulphate groups per disaccharide unit). The repeating unit of heparin consists of D-glucuronic acid (or L-iduronic acid) linked $\beta(1-4)$ to D-glucosamine (as stated previously for dermatan sulphate, since iduronate is of the L-configuration, the bond between iduronic acid and hexosamine is designated $\alpha$ rather than $\beta$). The disaccharides are linked glucosamine-$\beta(1,4)$-[hexuronic acid $\alpha/\beta$-$(1,4)$-glucosamine]$_n$-$\beta(1,4)$-hexuronic acid. The amino group on the glucosamine may be either sulphated or acetylated, also the C-6 carbon on this residue is generally sulphated. The iduronic acid residues may also be sulphated at C-2. Other reports have indicated that occasionally the C-3 position of D-glucosamine may also be sulphated. The molecular weight of heparin ranges from 6 to 25 kd.

### 1.1.5 *Heparan sulphate*

Heparan sulphate is very closely related to heparin, in that it consists of the same constituent monosaccharide residues, linked to one another in the same fashion as heparin. It varies from heparin, however, in that it contains higher proportions of glucuronic acid, is less *O*-sulphated and more *N*-acetylated. Moreover, heparan sulphate is less sulphated than heparin, having between 0.4 and 2 sulphate groups per disaccharide, so that many disaccharides may be unsulphated.

All of the glycosaminoglycans mentioned are found covalently bound to protein. The linkage to protein in the polyuronides involves an *O*-glycosidic bond between serine in the core protein and xylose found at the so-called 'linkage region' of the glycosaminoglycan (see *Figure 9*). Details of the structure and chemistry of the linkage of glycosaminoglycan to protein are excellently reviewed by Rodén (2).

### 1.2 **Proteoglycan general structure**

The best understood, most extensively examined group of proteoglycans are the high molecular weight aggregating proteoglycans from hyaline cartilage. This class of molecules have molecular weights ranging from 1 to 4 $\times$ $10^6$, however, only a small proportion of the molecule (5−15% dependent on species and age) is accounted for by protein. Although the protein contributes only a small fraction of the molecular weight, it contains both structural and functional domains essential to the function of the whole molecule. The domains found in cartilage proteoglycans are the hyaluronic acid-binding region, the keratan sulphate-rich region and the chondroitin sulphate-rich region. The hyaluronic acid-binding region is a globular portion of the core protein, having a molecular weight of about 65 000. It is highly substituted with *N*-linked oligosaccharides but only perhaps one or two glycosaminoglycan chains. This domain is responsible for the binding of proteoglycans to hyaluronic acid. The interaction is non-covalent, highly specific for hyaluronic acid and requires a strand of hyaluronate at least ten

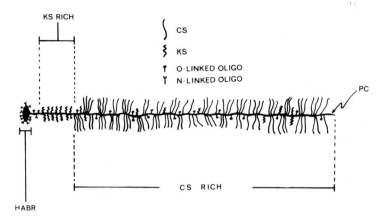

**Figure 1.** A schematic representation of the aggregating proteoglycans found in hyaline cartilage. Such proteoglycan monomers can associate via the hyaluronic acid-binding region (HABR) with hyaluronic acid to form huge multimolecular aggregates. CS: chondroitin sulphate; KS: keratan sulphate; *O*-linked oligo: *O*-linked oligosaccharide; *N*-linked oligo: *N*-linked oligosaccharide; HABR: hyaluronic acid-binding region; KS-rich: the keratan sulphate-rich region: CS-rich: the chondroitin sulphate-rich region; PC: protein core.

monosaccharides in length for binding. Another domain of the proteoglycan core protein is the keratan sulphate-rich region, which has a molecular weight of about 30 000. This region contains about 60% of the total keratan sulphate and about 10% of the chondroitin sulphate. The largest domain of the proteoglycan, however, is the chondroitin sulphate-rich region which represents around 60% of the total protein and has bound to it about 90% of the total chondroitin sulphate and 40% of the total keratan sulphate. A schematic representation of cartilage proteoglycan structure is shown in *Figure 1*.

A review of the detailed structure of the various proteoglycans of connective tissue is beyond the scope of this chapter, however, the subject is excellently reviewed by Heinegård and Paulsson (3) and Hardingham (4).

## 2. PROTEOGLYCAN EXTRACTION AND PURIFICATION

### 2.1 **Extraction**

The principal function of proteoglycans involves the retention of water within an extracellular matrix. In cartilage, proteoglycans distribute mechanical forces exerted on bone during locomotion by a hydraulic effect. This property of cartilage proteoglycans is due to the swelling pressure which they can exert when entrapped within the collagen network of the cartilage extracellular matrix. The swelling pressure arises as a result of a Donnan osmotic effect due to the high concentration of carboxyl and sulphate ester groups on the constituent glycosaminoglycans. To achieve this and various other biological effects, proteoglycans must be firmly maintained within the extracellular matrix. This may be achieved by specific interactions with components of the extracellular matrix as is found in tendon, where a dermatan sulphate proteoglycan specifically associates with the d-band of collagen (a similar interaction is found in cartilage, although it is not known whether the proteoglycan involved is a dermatan sulphate proteoglycan). Furthermore, the ability of many proteoglycans to form huge multimolecular aggregates (particularly cartilage proteoglycans) serves to entrap them within the dense insoluble collagen network of connective tissue matrices. Although such entrapment is beneficial for connective tissue function, it has been a great problem for biochemists interested in examining the structure of proteoglycans found in these tissues.

In the past, biochemical examination of the polysaccharide component of connective tissue involved the extraction of the tissue by alkali treatment or proteolytic digestion. As a result it was many years before it was realized that such glycosaminoglycans were found covalently attached to protein *in vivo*. Later attempts to extract intact proteoglycans involved procedures such as powdering at low temperature followed by neutral salt extraction. This did not cleave the glycosaminoglycan *O*-glycosidic bond to protein and the proteoglycans (then known as chondromucoproteins or mucopolysaccharides) isolated by such procedures were of considerably larger hydrodynamic size than the glycosaminoglycans found after alkali extraction. It has been shown, however, that such low temperature milling does produce extensive protein core cleavage and, as a result, has been abandoned as an extraction technique.

More recently the use of chaotropic agents such as $CaCl_2$, $MgCl_2$, $LaCl_3$ and guanidine.HCl have been used to extract proteoglycans intact and in relatively high yield. In particular, guanidine.HCl has achieved almost universal acceptance as an extractant for proteoglycan since it offers several advantages over the other chaotropes, particularly its ability to extract proteoglycans efficiently over a relatively wide range of concentrations and pH. At concentrations around 4 M, guanidine.HCl will dissociate proteoglycan aggregate structures and allow the proteoglycan monomers, link proteins and hyaluronic acid to diffuse out of tissues. Dialysis to more physiological salt concentrations allows almost complete re-assembly of the proteoglycan aggregates. In the presence of protease inhibitors and at a pH around 6, degradation of proteoglycan is not detectable.

Protease inhibitors which may be included in the extraction mixtures are (commonly): 100 mM 6-aminocaproic acid to inhibit cathepsin D; 5 mM benzamidine.HCl, which inhibits enzymes with activities similar to trypsin; 10 mM EDTA to inhibit metalloproteinases by chelating metal ions; 10 mM *N*-ethylmaleimide which inhibits both thiol proteases and also prevents disulphide exchanges (reactions common in such denaturing solvents); 1 mM phenylmethylsulphonyl fluoride (PMSF), a general inhibitor of serine proteases. Agents such as leupeptin, pepstatin, and soybean trypsin inhibitor may also be included in such protease inhibitor 'cocktails'.

Such mixtures of inhibitors have been included in extraction solvents as a prophylactic measure and the effect of individual components has not been determined empirically. It is not at all certain whether all these compounds are required to prevent proteolytic cleavage. It is, however, prudent to include protease inhibitors during all extraction and dialysis procedures, and perhaps include other inhibitors if a particular protease is thought to be active in the connective tissue to be examined.

### 2.1.1 *Dissociative extraction*

*Reagents:*

(A) 4 M Guanidine.HCl/0.05 M sodium acetate (pH 5.8) containing: 100 mM 6-aminocaproic acid/10 mM EDTA/5 mM benzamidine.HCl/10 mM *N*-ethylmaleimide/0.4 mM pepstatin/1 mM PMSF/soybean trypsin inhibitor (1 $\mu$g/ml) as protease inhibitors.

Dissolve 95.53 g of guanidine.HCl, 1.025 g of sodium acetate, 0.931 g of Na$_2$EDTA, 3.28 g of 6-aminocaproic acid, 0.196 g of benzamidine.HCl, 0.313 g of *N*-ethyl-maleimide and 250 $\mu$g of soybean trypsin inhibitor in distilled water, adjust to pH 5.8 with acetic acid and dilute to a final volume of 249 ml with distilled water. Just before use add PMSF (44 mg in 0.5 ml methanol) and pepstatin (43 mg in 0.5 ml methanol).

*Method:*

Tissue is maintained cool and moist by preparing the tissue in a Petri dish maintained on ice.

(i)     Finely slice the tissue using sharp scalpels into pieces no thicker than 0.2 – 0.5 mm and smaller if possible.

(ii)    Place the sliced tissue into suitably chilled extractant (reagent A) and stir for periods of up to 48 h at 4°C. For dense connective tissues such as cartilage, 10 volumes of reagent A are required. For looser connective tissues, e.g. aorta, basement membrane, corneal stroma, etc. about five volumes of reagent A is sufficient for optimal extraction.

(iii)   Residual tissue may be separated from the extract by centrifugation. Depending upon the density and size of the tissue the conditions of centrifugation to separate tissue from extract will vary. Generally, however, it is not necessary to exceed 20 000 *g* for 20 min at 4°C. Alternatively, the extract may be clarified by fil-tration through glass wool.

N.B. For optimal extraction of proteoglycan, the preparation of tissue is of paramount importance. For loose connective tissue matrices fine slicing is sufficient to obtain high extraction efficiency. In young animal cartilage, again, fine slicing is sufficient to allow extraction of a high proportion of the tissue proteoglycan. In dense connective tissues, such as elderly human articular cartilage, extraction from such slices produces poor yields. In particular, hyaluronic acid is extracted in very small percentages from such 'chip' extracts, giving the false impression that when dialysed to associative conditions, the proteoglycans contained in such extracts are less able to form aggregates than normal. It has been shown by Bayliss *et al.* (6) that microtomy of mature human cartilage is required to produce sufficiently thin slices of cartilage to allow high efficiency extraction. It is important, therefore, to determine the optimum tissue preparation which will allow maximum extraction of proteoglycan for each particular tissue examined.

## 2.1.2 *Associative extraction*

Although dissociative extraction is required to obtain high yields of proteoglycans from most connective tissues, it has been demonstrated that in loose, highly hydrated con-nective tissues such as the rat chondrosarcoma, considerable amounts of proteoglycan can be extracted with solvents which are not capable of dissociating proteoglycan aggre-gates [e.g. 0.5 M guanidine.HCl/0.05 M sodium acetate (pH 5.8)]. Such procedures are referred to as associative extraction techniques. Although used principally for the extraction of loose connective tissue, all connective tissues contain a proportion of proteo-glycans which may be extracted with these solvents. In articular cartilage, the proteo-glycans extracted by associative solvents largely represent those in the tissue which did not interact with hyaluronic acid. This non-interaction may be due either to an inability of such molecules to interact with hyaluronic acid (non-aggregating proteoglycans), or to the steric denial of access of aggregating proteoglycans to available hyaluronic acid. From loose connective tissues, it is possible with such associative solvents, to extract whole, intact proteoglycan aggregates. Such associatively extracted proteoglycan aggregates are often preferable to aggregate preparations prepared by dissociative extraction and subsequent dialysis to associative conditions. The associatively extracted aggregates represent a proportion of the proteoglycan population existing as aggregates in the tissue. A dissociatively extracted, reconstituted aggregate represents a randomization of all the proteoglycans extracted from the tissue and their interaction with a randomized population of hyaluronic acid.

*Reagents:*

(A) 0.5 M Guanidine.HCl/0.05 M sodium acetate (pH 5.8) containing the same protease inhibitors used in reagent A (Section 2.1.1).

Dissolve 11.948 g of guanidine.HCl, 1.025 g of sodium acetate, 3.28 g of 6-aminocaproic acid, 0.931 g of $Na_2EDTA$, 0.196 g of benzamidine.HCl, 0.313 g of $N$-ethylmaleimide and 250 $\mu$g of soybean trypsin inhibitor in distilled water. Adjust to pH 5.8 with acetic acid and the final volume to 249 ml with distilled water. Just before use add PMSF (44 mg in 0.5 ml methanol) and pepstatin (43 mg in 0.5 ml methanol).

*Method:*

Prepare tissue as indicated for dissociative extraction, bearing in mind the comments about the effects of tissue preparation on extraction efficiency. Perform the extraction at $0-4°C$ with constant stirring for periods of up to 48 h. Residual tissues may be separated from the extract by centrifugation or filtration as stated in Section 2.1.1 on dissociative extraction.

### 2.1.3 *Extraction with detergents*

The use of detergents to extract proteoglycans has become increasingly useful in the examination of proteoglycan biosynthesis in connective tissue cell culture. Detergent extraction is particularly suitable for extraction of intracellular proteoglycans and also for proteoglycans such as hepatocyte heparan sulphate proteoglycans which are intimately associated with plasma membranes. It would appear in many cases that the inclusion of detergent in high salt extraction buffers has a synergistic effect on extraction efficiency.

Detergents used in the extraction of proteoglycans include deoxycholate, Triton X-100 and the zwitterionic detergent, zwittergent ($N$-dodecyl-$N,N$-dimethyl-3-amino-1-propanesulphonic acid). Unfortunately, removal of such detergents is difficult and in most cases treatment of proteoglycan by these agents results in the loss of the ability to bind to hyaluronic acid. More recently, the detergent CHAPS (a zwitterionic derivative of cholic acid) has been used to solubilize intracellular proteoglycans from cells maintained in culture and it has been shown that such treatment maintains a functional binding region (7).

*Reagents:*

(A) Dissolve Triton X-100 (5 g) in reagent A described for the dissociative extraction procedure (final volume 100 ml).

*Method:*

(i)    Add reagent A (5 ml) for 10 min at 4°C to 35 mm culture dishes containing connective tissue cells which have been maintained in monolayer culture.

(ii)   Remove this first extract by aspiration and store at −20°C.

N.B. Detergent extraction may be used to extract whole tissues such as cartilage slices, basement membranes, etc. Such a protocol would generally involve an initial associative extraction (to remove non-aggregating proteoglycans) followed by a dissociative extraction (to remove the bulk of the proteoglycan). At this stage, extraction with detergent may be used to release a small proportion of pericellular or cell-associated proteoglycans from the tissue. Extraction with detergents is not commonly used in whole tissue extraction since the additional proteoglycan released from the tissue after 4 M guanidine.HCl extraction is usually small.

### 2.1.4 *Removal of 'non-extractable' (residual) proteoglycans*

None of the extraction procedures can either singly or in combination extract all the proteoglycans from tissue slices. The nature of the so-called 'non-extractable' proteoglycans is not certain, although the products of enzymic digestion of cartilage following extraction suggest that they share structural similarities with those extracted from the tissue (8). The non-extractable proteoglycans may be covalently linked to other components of the extracellular matrix, for example, the minor collagen species. Alternatively, it may be that in pericellular regions the collagen matrix may be so dense as to prevent the diffusion of proteoglycans during extraction.

Although it is technically difficult to extract all the proteoglycans intact from the tissue, in many cases one may need to estimate the amount of proteoglycan remaining within the tissue. This may be achieved by enzymic or chemical treatment of the residual cartilage and chemical estimation of the solubilized glycosaminoglycans.

### 2.1.4(i) *Enzymatic liberation of residual glycosaminoglycans*

*Reagent:*

(A) 0.5 M Tris-HCl/20 mM EDTA/20 mM cysteine HCl (pH 6.8) containing papain (13 mg%).

Dissolve 6 g of Tris, 77 mg of cysteine.HCl and 0.931 g of EDTA in distilled water, adjust the pH to 6.8 with HCl (final volume 100 ml). Add a suspension of papain (0.5 ml; available from Sigma Chemicals Ltd) to this solution. This gives a final papain concentration of 13 mg% (w/v).

*Method:*

(i)    Add the papain solution to the residual cartilage. Approximately 2 ml of solution is sufficient for up to 200 mg (wet weight) of cartilage.

(ii)   Incubate (in stoppered vessels) at 60−65°C for 16 h.

(iii)  If required, the reaction may be stopped by heating at 100°C for 10−20 min. If, however, one only wishes to assay for uronic acid or other carbohydrate components, such a step is not necessary.

2.1.4(ii) *Chemical liberation of residual glycosaminoglycans*

*Reagent:*

(A) 1 M sodium borohydride in 0.5 M NaOH.

Dissolve 0.378 g of sodium borohydride in 0.05 M NaOH (final volume 10 ml).

*Method:*

(i)    Add the alkaline borohydride reagent to the residual cartilage. Approximately 1 ml of reagent is sufficient to liberate the glycosaminoglycans from up to 100 mg (wet weight) of cartilage.

(ii)    Incubate the digests at 45°C, for 48 h in tightly stoppered vessels (open the lids periodically to let the hydrogen liberated by the reaction to escape).

(iii)    The reaction may be stopped by the addition of an equimolar amount of HCl. For the detailed chemistry of this reaction, see Rodén (2).

N.B. The borohydride is not essential for the cleavage of the O-glycosidic bond to protein. It does, however, protect the resultant glycosaminoglycan by reducing the xylose residue to xylitol and thereby prevents destruction of the carbohydrate chain by 'chain-peeling' reactions. Moreover, one can radioactively label the resultant glycosaminoglycans if required by using $NaB^3H_4$.

## 2.2 Proteoglycan purification by isopycnic CsCl density gradient centrifugation

When subjected to high centrifugal fields, caesium salts ($CsCl$, $CsBr$, $Cs_2SO_4$, etc.) spontaneously form concentration gradients. When cartilage extracts are centrifuged under these conditions, proteoglycans are recovered in the high density regions of the gradient, whereas proteins migrate to the least dense part. This phenomenon is due to the binding of large amounts of caesium ions to the negatively charged groups on the glycosaminoglycans. Proteins, however, bind considerably less caesium and hence have a much lower buoyant density.

This technique was first used to purify glycoproteins by Franek and Dunstone (9), was adapted by Sajdera and Hascall for proteoglycans (10) and has become universally accepted as a preparative method since it offers many advantages over other purification schemes:

(i)    it has a high capacity;

(ii)    it does not require gradient-forming apparatus;

(iii)    proteoglycans may be purified in an aggregated form;

(iv)    proteoglycan aggregates may be subsequently separated into their component parts (see Section 2.2.2) by recentrifuging in the presence of 4 M guanidine.HCl;

(v)    the separation is not dependent on size, hence all high-density molecules can be recovered free from high-molecular weight protein contaminants.

### 2.2.1 *Associative density gradient centrifugation*

This technique is used to separate proteoglycan aggregates from extracellular matrix proteins.

*Method:*

(i)    Dialyse the proteoglycan extract (containing <6 mg/ml uronic acid) against 0.5 M guanidine.HCl/sodium acetate (pH 5.8) containing protease inhibitors (this solution is reagent A, described in Section 2.1.2, on the associative extraction of proteoglycan).

(ii) Add solid CsCl to the extract to give the solution a final density of 1.5 g/ml. This may be achieved as follows: to 4 ml of extract add 3.4 g of CsCl which will produce a solution whose final volume is 5 ml and whose density is 1.5 g/ml.

(iii) Centrifuge the solution at 100 000 *g* for 48 h at 5−10°C. Centrifugation may be carried out in fixed angle, swing-out or vertical rotors, although vertical rotors are not particularly suitable for the centrifugation of analytical amounts of material (they are very useful, however, for large-scale preparations of proteoglycans).

N.B. The initial density ($\varrho_i$) used in this method is lower than stated in many other protocols. This has the following advantages: (i) it ensures that CsCl does not precipitate at the densest part of the gradient and (ii) it serves to compact proteoglycan at the bottom of the gradient. Different connective tissues contain proteoglycans of varying buoyant density and one may wish to alter the starting density to obtain optimum recovery. Hence one might choose a lower starting density for proteoglycans with a low buoyant density or higher starting densities for proteoglycans with relatively high buoyant density.

### 2.2.2 *Dissociative density gradient centrifugation*

This procedure is used to prepare monomeric proteoglycans, free from the link proteins and hyaluronic acid found in aggregate structures.

*Method:*

(i) Mix the A1 fraction (the fraction of the associative gradient of the highest density) with an equal volume of 7.5 M guanidine.HCl/0.05 M sodium acetate (pH 5.8) (containing the same protease inhibitors included in reagent A used for associative and dissociative extraction of proteoglycan in Sections 2.1.1 and 2.1.2).

(ii) Add solid CsCl to the required initial density as indicated in *Table 2*. Normally an initial density of 1.5 g/ml is suitable for the purposes of this separation.

**Table 2.** Preparation of dissociative density gradients

| (A) Density of starting material (g/ml) | (B) Required density (g/ml) | | | | |
|---|---|---|---|---|---|
| | 1.35 | 1.40 | 1.45 | 1.50 | 1.55 |
| 1.35 | 0.153 | 0.238 | 0.323 | 0.408 | 0.493 |
| 1.40 | 0.100 | 0.187 | 0.274 | 0.361 | 0.448 |
| 1.45 | 0.047 | 0.136 | 0.225 | 0.314 | 0.403 |
| 1.50 | − | 0.085 | 0.176 | 0.267 | 0.358 |
| 1.55 | − | 0.034 | 0.127 | 0.220 | 0.313 |
| 1.60 | − | − | 0.078 | 0.173 | 0.268 |
| 1.65 | − | − | 0.029 | 0.126 | 0.223 |
| 1.70 | − | − | − | 0.079 | 0.178 |

Column (A) refers to the density of the A1 fraction which is to be subsequently centrifuged under dissociative conditions. Columns (B) refer to the required starting density for the subsequent dissociative density gradient run. The figures in the Table refer to the amount of CsCl (in g/ml of solution) to be added to the A1 fraction (after addition of an equal volume of 7.5 M guanidine.HCl to it) to achieve the required starting density for the dissociative gradient.
N.B. The addition of 0.5 g CsCl/ml of solution produces an approximately 10% increase in final volume.

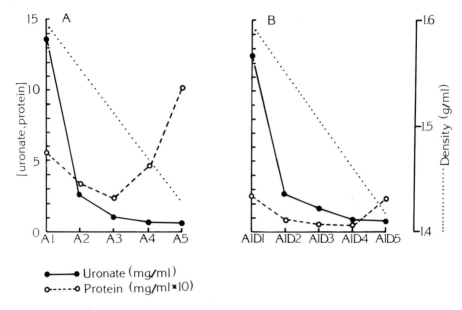

**Figure 2.** Isopycnic density gradient centrifugation of proteoglycans in CsCl. **A**. Associative density gradient centrifugation of bovine nasal cartilage extracts. The bulk of the proteoglycan is recovered in the densest (A1, A2) fractions. Collagen and non-collagenous matrix proteins are recovered in the least dense (A4, A5) fractions. **B**. Subsequent dissociative density gradient centrifugation of the A1 fraction from **A**. In this case the aggregate structure is disrupted into proteoglycan monomers, which are recovered in the densest (A1D1, A1D2) fractions, link protein (found in the A1D5 fraction) and hyaluronic acid which is distributed diffusely throughout the central portion of the gradient.

(iii)    Centrifuge the solution at 100 000 *g* for 48 h at 5−10°C. Proteoglycan distribution in associative and subsequent dissociative CsCl density gradient centrifugation is shown in *Figure 2*.

### 2.2.3 *Direct dissociative gradient centrifugation*

One may directly centrifuge dissociative (i.e. 4 M guanidine.HCl) extracts of tissues to separate proteoglycan monomers from the other components of proteoglycan aggregates (hyaluronic acid and link proteins) as well as all the other proteins of the extracellular matrix. This separation is essentially as outlined previously except that CsCl is added directly to the extract to the required density.

*Method:*

(i)    Add solid CsCl to the dissociative proteoglycan extract to the required starting density. Normally a density of 1.5 g/ml is suitable for most applications although densities as low as 1.35 g/ml may be used where the proteoglycans to be separated are of particularly low buoyant density (for example the proteoglycans from vascular connective tissue). To achieve a starting density of 1.5 g/ml add 6.52 g of CsCl to 10 ml of extract, which gives a final volume of 11.7 ml.

(ii)    Centrifuge the solution at 100 000 *g* for 48 h at 5−10°C.

### 2.2.4 *Fractionation of gradients*

There are four basic methods by which density gradients may be fractionated: (i) by upward displacement; (ii) by piercing and draining from the bottom of the tube; (iii) by aspiration from the top of the gradient and (iv) freezing the tubes and sawing into equal fractions (by volume). In general, when preparing bulk amounts of proteoglycan, upward displacement and downwards draining of the gradients are not particularly suitable since the proteoglycan forms a dense gel in the bottom of the tube. Aspiration of fractions from the top of the gradient is an adequate method of fractionation in this case. Freezing of the gradients rapidly in a methanol-dry ice bath and subsequent sawing, however, is perhaps the quickest and easiest method of fractionation of such gradients. When fractionating analytical amounts of proteoglycan any of the methods mentioned above are suitable and selection of any one procedure is largely a matter of personal preference.

## 2.3 **Proteoglycan purification by ion exchange chromatography**

Proteoglycans, due to the high concentration of carboxyl and sulphate ester groups, are highly negatively charged. As a result, they bind very tightly to anion exchangers and this property can be used to separate such molecules from other proteins which have a considerably lower charge density. An ion exchange technique for proteoglycan preparation using DEAE-cellulose as the anion exchanger was developed by Antonopoulos *et al.* (11). In this resin, the charged group is the positively charged amino residue of the diethylaminoethyl [$-OCH_2CH_2NH^+(CH_2CH_3)_2$] substituent on the cellulose matrix. The procedure is performed in the presence of 7 M urea which serves two main purposes, (i) it maintains the proteoglycans in a disaggregated form (in the absence of a chaotropic agent, cartilage proteoglycan aggregates do not bind particularly well to DEAE-cellulose, and (ii) it prevents the formation of colloidal suspensions which may occur on dialysis to associative conditions.

The urea used in this procedure should be as free of ionic contaminants as possible since at concentrations of 7 M these may interfere with the binding of molecules to the exchanger and their subsequent elution. Either one may use ultrapure urea (which is expensive) or alternatively one may remove the ionic contaminants (largely cyanates) from a concentrated urea solution by passing it through a column containing mixed-bed ion-exchange resin (Dowex-50W $H^+$ form and Dowex 1 $Cl^-$ form are suitable exchangers for this purpose). The urea may then be freeze-dried and used as required.

### 2.3.1 *Preparation of DEAE-cellulose*

*Reagents:*

(A) 7 M Urea/0.05 M Tris-HCl (pH 6.5).

Dissolve 210 g of urea and 3.03 g of Tris in distilled water and adjust the pH to 6.5 with HCl and to a final volume of 500 ml with distilled water.

(B) 7 M Urea/0.05 M Tris-HCl/0.15 M NaCl.

Dissolve 210 g of urea, 3.03 g of Tris and 4.38 g of NaCl in distilled water and adjust the pH to 6.5 with HCl. Adjust to a final volume of 500 ml with distilled water.

(C) 7 M Urea/0.05 M Tris-HCl/2 M NaCl.

Dissolve 210 g of urea, 3.03 g of Tris and 54.44 g of NaCl in distilled water and adjust the pH to 6.5 with HCl. Adjust to a final volume of 500 ml with distilled water.

(D) 0.5 M HCl.

Add 43 ml of concentrated HCl to 957 ml of distilled water.

*Method:*

(i)     Add DEAE-cellulose to distilled water (10 g per litre) and allow to swell. This normally takes between 16 and 24 h at room temperature.

(ii)    Transfer the slurry to a sintered glass funnel and wash with 50−100 volumes (i.e. 50−100 ml per ml ion exchange resin) distilled water.

(iii)   Resuspend the resin cake in about 10−20 volumes of 0.5 M HCl (reagent D) in a large measuring cylinder and allow the suspension to settle for 1−2 h at room temperature.

(iv)    Carefully decant off the supernatant (this removes finings from the exchanger). Transfer the slurry to a sintered glass funnel and wash with distilled water until the pH of the washes approaches 8.0.

(v)     Resuspend the cake in reagent A [7 M urea/0.05 M Tris-HCl (pH 6.5), about 10−20 volumes] and allow to stand for 1−2 h at room temperature. Carefully decant off the supernatant (to remove any remaining finings) and dilute the packed slurry 1:1 (v/v) with reagent A.

## 2.3.2 *Selection of column*

Ion-exchange chromatography is generally performed in relatively short, broad columns in contrast with gel permation chromatography. The bed supports should be selected such that with continued use they will not become clogged. Hence glass wool and coarse glass sinters are not generally recommended for long-term, prolonged use. Many manufacturers supply columns suitable for ion-exchange chromatography and selection is largely a matter of personal choice.

Generally one need not use columns of greater than 20 cm in length. Since one may load about 2 mg of proteoglycan per ml of DEAE-cellulose, to load 100 mg of proteoglycan, a column of 20 cm in length and 3 cm in diameter containing about 125 ml of DEAE-cellulose is sufficient.

## 2.3.3 *Packing the column*

(i)     Degas the DEAE-cellulose slurry under vacuum for about 20 min.

(ii)    Mount the column vertically on a suitable stand or rack and flush the afferent tubing (and end-pieces if being used) with buffer to release any trapped air, then clamp and close the outlet.

(iii)   Pour the exchanger suspension into the column either down a glass rod or against the side of the column, taking care to avoid forming bubbles. Open the column outlet and allow the exchanger to settle into the column. The column should be packed at a rate in excess of that to be used for the separation.

(iv)    When all the exchanger has been packed into the column, overlay with reagent A to a height of 2−3 cm. The top of the column may then be connected to a reservoir containing reagent A.

(v) Adjust the column outlet such that the operating pressure is about 1 cm buffer/cm height of exchanger (i.e. for a column containing 20 cm of exchanger, the difference in height between the level of buffer in the reservoir and the column outlet is 20 cm).

(vi) Run the buffer through the column until the conductivity (measured using a conductivity meter) and pH of the eluant and effluent are equal. The column is now correctly equilibrated and ready for use.

### 2.3.4 *Sample preparation*

Guanidine.HCl extracts of connective tissue must be adjusted to 7 M urea/0.05 M Tris-HCl (pH 6.5) before application to the column. This may be achieved by dialysis into this buffer (with several changes). It is advisable to then concentrate this extract, which may conveniently be achieved by vacuum dialysis. Alternatively, one may dilute the 4 M guanidine.HCl abstract and concentrate by vacuum dialysis, dilute with the 7 M urea buffer and continue to vacuum dialyse, repeating the dilution step until one has completely exchanged the guanidine.HCl for 7 M urea. The sample is then ready for application to the ion-exchange column.

### 2.3.5 *Sample application*

The maximum flow rate which one can use for separations is dictated by the cross-sectional area of the column such that wider columns may be run at a higher flow rate than narrow ones. Optimal flow rates for any particular separation, however, must be determined empirically.

There are three ways by which samples may be applied:

(i) Using an adaptor; this is normally the most convenient way to apply samples to ion-exchange columns.

(ii) Application under the buffer; samples may be made more dense by the addition of an uncharged compound such as glucose. Such a dense solution may then be carefully layered onto the surface of the column using a hypodermic syringe after closing the column outlet.

(iii) Application onto a drained bed; on the face of it, this method appears to be the most simple; however, it is perhaps the most difficult to perform reproducibly. Disconnect the column from the reservoir and allow the eluant to drain to the bed surface. Clamp the column outlet (and/or switch off the pump) then carefully layer the sample onto the bed surface. This application must be even, gradual and careful in order to prevent distortion or disturbance of the bed surface, which would result in bands skewing on elution from the exchanger.

Under normal circumstances, the third technique is normally adopted, and with practice may be as reproducible and efficient as any other sample application method. After sample application either under the buffer or onto a drained bed, the column is reconnected to the buffer reservoir and the column outlets opened, and pump (if used) switched on.

### 2.3.6 *Elution*

There are two forms of elution which can be used for the separation of proteoglycans from other connective tissue matrix proteins: first, stepwise elution and second, gradient elution.

(a) *Stepwise elution.* After sample application, the column is eluted with reagent A until all the proteins which do not bind to the column (or those which bind only weakly) have eluted from the column. Normally two or three bed volumes is sufficient for this purpose. Following this initial elution step, elute the column with reagent B [7 M urea/0.15 M NaCl/0.05 M Tris-HCl (pH 6.5)]. This step elutes the remainder of the non-proteoglycan connective tissue matrix proteins from the column. Elution with this buffer is continued until no material elutes from the column, again normally two or three bed volumes is required. Finally, elute the column with reagent C [7 M urea/2 M NaCl/0.05 M Tris-HCl (pH 6.5)]. Approximately 90% of the proteoglycan elutes from the column in this buffer and, as before, two or three column volumes is sufficient to achieve complete removal of proteoglycan from the ion-exchange matrix.

This particular procedure is useful for large-scale preparations of proteoglycans. One may wish, however, to separate sub-species of proteoglycan from extracts of connective tissue. This may be achieved by eluting the ion-exchange column with a linear salt gradient, as in (b) below.

(b) *Gradient elution.* After sample application, elute the column with two or three volumes of reagent A. As previously stated, under these conditions proteins which do not bind, or only bind weakly will be eluted from the column. Apply a linear NaCl gradient from 0 to 2 M to the column as outlined below.

### 2.3.7 *Preparation of gradient*

*Method:*

(i)     Using a suitable gradient former, pour into the compartment for the dense solution 10−25 bed volumes of reagent C. Into the compartment for the light solution, pour an equal volume of reagent A (details of the composition of reagents A and C are given in Section 2.3.1).

(ii)    Using a magnetic stirrer, mix the solution in the compartment containing reagent A.

(iii)   Connect the ion-exchange column to the compartment containing reagent A.

(iv)    Open the connection between the compartments containing A and C.

Following elution of the ion-exchange column with such a linear salt gradient, continue elution with two or three bed volumes of reagent C to ensure complete elution of proteoglycans from the column.

N.B. Many new matrix supports have been introduced for ion-exchange chromatography. Of particular interest are DEAE-Sephacel, a microcrystalline bead-formed exchanger whose support matrix is cellulose. As a result of its bead structure, it offers greater resolution than existing DEAE-cellulose exchangers. This may be of importance if one wishes to separate proteoglycans having only narrowly differing elution characteristics on DEAE-cellulose. Furthermore, a DEAE exchanger using a gel permeation matrix as support [DEAE-(Sepharose CL-6B)] has been developed which has better flow characteristics than any of the ion-exchangers previously mentioned. Moreover, such a matrix affords a certain element of gel permeation to the separation, which may be of use when preparing low molecular weight proteoglycans of a size fractionable by Sepharose 6B.

# 3. PROTEOGLYCAN ANALYTICAL TECHNIQUES

## 3.1 **Gel permeation chromatography**

Gel permeation chromatography of proteoglycans has become the method of choice for the routine laboratory analysis of their hydrodynamic size. Such separations may be achieved on large-pore matrices based on either agarose or dextran gels or alternatively on controlled pore glass supports. Although all these matrices have been successfully used to fractionate proteoglycans, the agarose-based gels have become most widely accepted for this purpose. The most commonly used agarose gels are the Sepharoses (produced by Pharmacia, Uppsala, Sweden) and will be described in the following section.

*Reagents:*

(A) Associative buffer. 0.5 M Sodium acetate (pH 6.8).

Dissolve 41 g of sodium acetate in distilled water. Adjust the pH to 6.8 with acetic acid and the final volume to 1 l with distilled water.

(B) Dissociative buffer. 2 M Guanidine.HCl/0.5 M sodium acetate (pH 6.8).

Dissolve 190 g of guanidine.HCl and 41 g of sodium acetate in distilled water. Adjust the pH to 6.8 with acetic acid and the final volume to 1 l with distilled water.

Before use, filter all buffers through a Millipore filter (0.45 $\mu$) and degas under vacuum with gentle heating for about 20 min.

### 3.1.1 *Preparation of gel matrices*

(i)   Transfer the Sepharose slurry to a sintered glass funnel and wash with $10-20$ volumes of reagent A, to equilibrate the gel and remove the Merthiolate present in the gel as a preservative. Sepharose is supplied preswollen so no swelling of the matrix is required.

(ii)  Suspend the gel cake in 10 volumes of reagent A, and allow to settle in a suitable measuring cylinder.

(iii) Carefully decant off the supernatant to remove finings, then suspend the gel in an equal volume of reagent A.

(iv)  Degas the gel suspension under vacuum with gentle heating.

N.B. The gel matrix prepared as described has been equilibrated for elution under associative conditions. To prepare the gel for elution under dissociative conditions, all equilibration steps should be performed using reagent B instead of reagent A.

### 3.1.2 *Selection of column*

Gel permeation chromatography should be performed in long narrow columns, since the resolution of macromolecules increases in proportion to the square root of column length. Broad columns produce an unwanted dilution of sample, however, excessively narrow columns are difficult to pack and may suffer from cohesive wall effects. As a general rule, however, the length of the column should be dictated by the optimal resolution required for the separation, whereas the diameter should be selected according to the quantity of sample. One should ensure that the column outlet deadspace is as small as possible to prevent the remixing of resolved compounds. A suitably sized column would be 100 cm in length and $0.8-1.5$ cm internal diameter. A column this size would be capable of resolving up to 400 $\mu$g (as uronic acid) of applied proteoglycan.

### 3.1.3 *Column packing*

(i) Fill the column with degassed buffer (reagent A or B depending upon whether the column is to be eluted under associative or dissociative conditions). Flush any air from the column effluent end-pieces and clamp the column closed.

(ii) Add the gel slurry to the column. Preferably, this should be done in one step, using a reservoir attached to the column. If such a packing reservoir is not available, slurry may be periodically added using a Pasteur pipette.

(iii) Open the column outlet and allow the gel to settle in the column. Ideally, the flow rate during packing should be slightly faster than that to be used during chromatographic separations.

(iv) When the gel has completely packed down, connect the top of the column to a suitable reservoir and reduce the column flow rate to that to be used for separative runs. Allow two bed volumes of buffer to flow through the column before use.

(v) Before samples are chromatographed, it is advisable to 'precondition' the column by applying a proteoglycan sample ($200-300$ $\mu$g uronic acid) to prevent subsequent non-specific binding of samples to the column. Following preconditioning, allow about two bed volumes of buffer to elute from the column before use.

(vi) A suitable flow rate for chromatography on Sepharose columns is between about 10 and 15 ml/cm$^2$/h.

### 3.1.4 *Sample application*

Samples may be applied to the column in any of the ways outlined in the section on ion-exchange chromatography of proteoglycans. It is common, however, to apply the sample onto a drained column bed.

### 3.1.5 *Gel chromatography of proteoglycans on Sepharose gels*

Sepharose 2B (or CL-2B) is an agarose-based gel permeation matrix having an approximate agarose concentration of 2%. It has a very high exclusion limit, such that monomeric proteoglycans of the high molecular weight cartilage type will be included on these columns. Proteoglycans which are interacting with hyaluronic acid will, however, be of sufficiently large molecular size to be excluded from the column. Thus, chromatography of proteoglycans on columns of Sepharose 2B can be used to assess the proportion of such molecules which can interact with hyaluronic acid. Chromatography of proteoglycans on columns of Sepharose 2B eluted under dissociative conditions may be used to assess the relative molecular size of such preparations. Typical elution profiles of proteoglycan on Sepharose 2B are given in *Figure 3*.

Although cartilage-type proteoglycans are included on columns of Sepharose 2B, they are completely excluded from columns of Sepharose 4B (or CL-4B) even when eluted under dissociative conditions. Such columns may be used, however, to examine breakdown products of such high molecular weight proteoglycans which may be partly included on Sepharose 4B columns. Moreover, lower molecular weight proteoglycans such as those found in bone and skin are well included on these columns.

Sepharose 6B or (CL-6B) has an agarose concentration of about 6% and can therefore fractionate proteoglycans too small to be effectively fractionated on either Sepharose

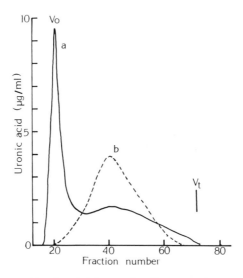

**Figure 3.** Gel permeation centrifugation of pig laryngeal cartilage proteoglycans on Sepharose 2B. **(a)** Proteoglycan aggregates eluted under associative conditions (—). Proteoglycans interacting with hyaluronic acid are excluded from columns of Sepharose 2B and hence elute at the $V_0$. A proportion of proteoglycan preparations will not interact with hyaluronic acid (the so-called 'non-aggregating' proteoglycans) and hence are included on such columns. **(b)** Proteoglycan preparations chromatographed on Sepharose 2B columns eluted under dissociative conditions (- - -) are disaggregated and hence are totally included on such columns.

**Table 3.** Chromatography of various proteoglycans on Sepharose gel permeation matrices

| | | Sepharose 2B | | Sepharose 4B DISS | Sepharose 6B DISS |
|---|---|---|---|---|---|
| | | ASS | DISS | | |
| Skeletal proteoglycans | Cartilage[b] | E | I[a] | E | E |
| | Tendon[b] | E | I[a] | E | E |
| | Bone | I | I | I[a] | E |
| | Meniscus[b] | E | I[a] | E | E |
| | Intervertebral disc[b] | E | I[a] | E/I | E |
| Non-skeletal proteoglycans | Follicular fluid | I[a] | I[a] | E | E |
| | Skin | I | I | I[a] | E |
| | Aorta[b] | E | I[a] | E | E |
| | Corneal stroma (KS) | HI | HI | HI | I[a] |
| | Lung parenchyma | HI | HI | I[a] | E |
| | Glial cell[b] | E | I[a] | E | E |
| | Epithelial cell HS-PG | I | I | I[a] | E |

E indicates that the proteoglycans are excluded from the matrix; I indicates that the proteoglycans are included in the matrix; HI indicates that the proteoglycans are highly included in the matrix. ASS indicates that the column is eluted under associative conditions, DISS indicates elution under dissociative conditions.
[a]Indicates the most suitable gel matrix for the fractionation of a particular proteoglycan species.
[b]These proteoglycans are excluded from columns of Sepharose 2B, eluted under associative conditions, when aggregated with hyaluronic acid.

2B or 4B. This matrix may also be used to fractionate glycosaminoglycans in addition to small proteoglycans possessing only a few glycosaminoglycan chains, such as found in corneal stroma. *Table 3* shows suitable matrices for the fractionation of proteoglycans from various tissues.

N.B. When comparing relative proteoglycan sizes, it is common to refer to their '$K_{av}$' on a particular column. This term is a distribution coefficient and is related to the proportion of the column matrix which is available to a particulate solute species. The $K_{av}$ may be calculated from the following expression:

$$K_{av} = \frac{V_e - V_0}{V_t - V_0}$$

where $V_e$ is the elution volume of the solute in question; $V_0$ is the void volume of the column (i.e. that volume in which very large molecules which are excluded from the gel will elute); $V_t$ is the total volume of the column (i.e. that volume in which very small molecules, such as water, which are completely included in the gel, will elute).

## 3.2 Determination of the degree of link stabilization of proteoglycan aggregates

This assay is based on the observation that the interaction of proteoglycan with hyaluronic acid (HA) is freely reversible under physiological conditions whereas the ternary complex formed between proteoglycan, link protein and hyaluronic acid is effectively irreversible (12,13). Thus, by including a molar excess of oligosaccharides of HA about 10 – 30 monosaccharides in length to a proteoglycan preparation those proteoglycans participating in non-link stable interactions with HA will be competed off the HA by the HA oligosaccharides (and hence be included on columns of Sepharose 2B, eluted under associative conditions) but link-stabilized aggregates will remain intact (and be excluded from such Sepharose 2B columns).

### 3.2.1 *Preparation of oligosaccharides of hyaluronic acid (14)*

*Reagents:*

(A) 0.15 M NaCl/0.1 M sodium acetate (pH 5).
    Dissolve 0.888 g of NaCl and 82 mg of sodium acetate in distilled water and adjust to pH 5 with acetic acid and to a final volume of 10 ml with distilled water.

*Method:*

(i)     Dissolve HA (25 mg) in reagent A (2 ml).
(ii)    Add bovine testicular hyaluronidase (50 μg) and incubate for 3 h at 37°C.
(iii)   Heat the solution for 5 min at 100°C then cool to room temperature.
(iv)    Chromatograph the digest on a column of Sepharose G-50 (superfine) eluted with 0.25 M pyridinium acetate pH 6.5 (essentially as outlined in Section 3.1).
(v)     Retain fractions having a $K_{av}$ between 0.3 and 0.45 (see Section 3.1.5), rotary evaporate these fractions and redissolve in water (~10 ml). Rotary evaporate again to dryness (continue repeated evaporations until no trace of pyridine may be detected).
(vi)    When free from pyridine, redissolve the oligosaccharides in water (1 ml).

### 3.2.2 *Determination of link stability*

(i)     Dialyse the proteoglycan preparation into associative conditions [i.e. 0.5 M guanidine.HCl, 0.05 M acetate (pH 5.8) which is reagent A used in Section 2.1.2].

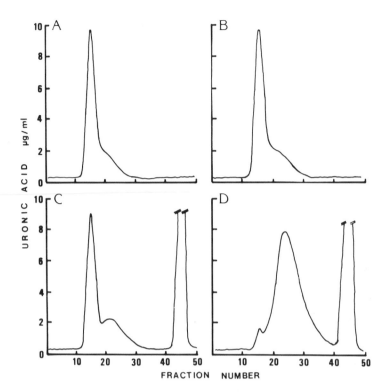

**Figure 4.** Demonstration of link stabilization of pig laryngeal cartilage proteoglycan aggregates. Sepharose 2B chromatographs of proteoglycan-hyaluronate complexes eluted under associative conditions in the presence (**A**) and absence (**B**) of link protein. Aliquots of the material chromatographed on **A** and **B** were incubated with an excess of hyaluronic acid oligosaccharides as indicated in Section 3.2.2 and chromatographed on columns of Sepharose 2B eluted under associative conditions, giving rise to chromatographs **C** and **D** respectively. Those proteoglycans participating in link-stabilized interactions with hyaluronic acid are resistant to dissociation with hyaluronic acid oligosaccharides and are excluded from columns of Sepharose 2B (**C**). Those not participating in link-stabilized interactions are dissociated in the presence of hyaluronic acid oligosaccharides (by competitive inhibition of the interaction) and are included on Sepharose 2B columns (**D**).

(ii)    Add 10 mg HA oligosaccharides per mg of proteoglycan (both measured as uronic acid equivalents).

(iii)   Incubate the mixture at 4°C for 16 h.

(iv)   Chromatograph the mixture on a column of Sepharose 2B eluted under associative conditions, as outlined in Section 3.1.

(v)    Assay the eluant fractions for uronic acid (if using analytical amounts of proteoglycan) or measure $^{35}S$ (if using biosynthetically radiolabelled [$^{35}S$]proteoglycans). As shown in *Figure 4* only those proteoglycans participating in link stabilized interactions with HA will be excluded from columns of Sepharose 2B. The degree of link stabilization can be determined by measurement of the area of the included and excluded peaks and may be expressed as follows:

$$\% \text{ link stabilization} = \frac{(\text{area of excluded peak}) \times 100}{(\text{area of excluded peak}) + (\text{area of included peak})}$$

### 3.3 **Preparation of proteoglycan hyaluronic acid binding region and link protein**

The hyaluronic acid-binding region (HABR) and link protein (LP) are important components of proteoglycan aggregates in hyaline cartilage. The finding that many tissues from a wide range of species possess both HABR and LP closely related to those from cartilage, both structurally and immunologically, would suggest that such molecules are highly conserved and are of great importance to the structural integrity of many connective tissues.

The HABR is a globular domain of proteoglycan core protein which specifically interacts with hyaluronic acid allowing the formation of multimolecular aggregates. It has a high affinity ($K_d$ ~2 × 10$^{-8}$ M) for hyaluronic acid and forms a ternary complex with it and LP, which under normal physiological conditions is not reversible and cannot be competitively inhibited by HA oligosaccharides. The preparation of HABR and LP in highly pure form has allowed these structures to be studied *in vivo* and *in vitro* using immunological techniques.

The presence of LP in the ternary complex protects the HABR from limited proteolytic digestion (15); however, the rest of the proteoglycan molecule is highly degraded. The large molecular size of the HA-LP-HABR complex allows it to be easily separated from the degraded proteoglycan fragments by preparative gel chromatography under associative conditions. The LP and HABR may then be separated from one another by chromatography under dissociative conditions.

*Reagents:*

(A) 0.1 M Tris/0.05 M sodium acetate (pH 7.3).

Dissolve 12.1 g of Tris and 4.1 g of sodium acetate in distilled water, adjust the pH with acetic acid to 7.3 and the final volume to 1 l with distilled water.

(B) 4 M Guanidine.HCl/0.05 M sodium acetate/1 mM EDTA (pH 5.8).

Dissolve 380 g of guanidine.HCl, 4.1 g of sodium acetate and 0.372 g of EDTA in distilled water, adjust the pH with acetic acid to 5.8 and the final volume to 1 l with distilled water. Filter (Millipore: 0.45 $\mu$) and degas thoroughly before use.

*Method:*

(i)    Prepare proteoglycan aggregates (A1 fraction) as outlined in Section 2.2.1. Freeze-dry the preparation and redissolve in reagent A.

(ii)   Add chondroitinase ABC (3.5 units/g proteoglycan) to the aggregate preparation and allow to digest for 50 min at 37°C.

(iii)  Add trypsin (diphenylcarbamoyl chloride treated) to the solution (2 mg/g proteoglycan) and allow to digest for 6 h at 37°C.

(iv)   Concentrate the digest to about one-third of its volume by partial freeze-drying then chromatograph on a column of Sepharose CL-6B eluted under associative conditions (as outlined in Section 3.1). A typical profile is shown in *Figure 5*.

(v)    A protein-rich fraction (peak 'A', *Figure 5*) is excluded from the CL-6B column. This contains the HA-LP-HABR complex. Pool the fractions containing peak 'A' and dialyse against distilled water and freeze-dry.

(vi)   Dissolve the freeze-dried sample in reagent B.

(vii)  Chromatograph the sample obtained from (vi) on a gel permeation column containing Sephacryl S-300 (prepared essentially by the methods outlined in Section 3.1), equilibrated and eluted with reagent B.

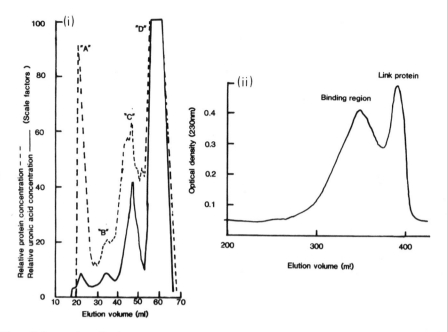

**Figure 5.** Preparation of hyaluronic acid-binding region and link protein. Proteoglycan aggregates are digested with chondroitinase ABC giving rise to a KS-protein core substituted with CS-'stubs'. Trypsin digestion of this material gives rise to a hyaluronic acid-HABR-link protein complex (A), a keratan sulphate-rich region (B), CS 'stub'-peptides (C), chondroitin sulphate disaccharides and small tryptic peptides (D). **(i)** Chromatography of chondroitinase ABC/trypsin-digested proteoglycan aggregates on a column of Sepharose 6B eluted under associative conditions ($V_0$ = 22.5 ml, $V_t$ = 54 ml). **(ii)** Chromatography on Sephacryl S-300 ($V_0$ = 245 ml, $V_t$ = 740 ml) eluted under dissociative conditions of peak A. The first peak contains HABR, the peak eluting after HABR contains largely link protein. Data courtesy of Mr D.G.Dunham and Dr T.E. Hardingham.

(viii)   As shown in *Figure 5*, HABR and LP are partially separated by chromatography on Sephacryl S-300. Pool the fractions containing (i) HABR and (ii) LP and rechromatograph on Sephacryl S-300 to further purify.

N.B. LP may be prepared from aggregate (A1) fractions without enzymic digestion. As described in Section 2.2, CsCl density gradient centrifugation of A1 preparations under dissociative conditions will separate the components of aggregate structures. Hence, the A1D5 fractions will be highly enriched in LP. Dialysis against reagent B and subsequent chromatography on Sephacryl S-300 as described in (vii) and (viii) above will yield a highly purified LP preparation.

This purification scheme was used for young porcine laryngeal cartilage (16). Tissue from other species or even tissue from older pigs may give qualitatively different chromatographic profiles for the digestion products. The HA-LP-HABR complex will, however, always behave in a chromatographically similar way under the conditions described.

## 3.4 Analytical ultracentrifugation of proteoglycans

### 3.4.1 *Sedimentation velocity*

This is the most commonly used method for centrifugal analysis of proteoglycan size.

119

Subjecting a homogeneous solution of proteoglycan to high centrifugal fields causes the solute particles to migrate away radially from the centre of rotation. Thus, a boundary (or boundaries, depending on the number of different species present) develops between the solvent from which the solute molecules have vacated and the solvent which still contains proteoglycan solute molecules. This method of determination of sedimentation coefficient is known as the moving boundary method. The rate of migration of this boundary (or boundaries) is directly proportional to the sedimentation rate of the solute particles which may be calculated from the equation:

$$S = \frac{dx}{dt} \times \frac{1}{\omega^2 r}$$

where $r$ = radial distance (cm), $t$ = time (s), $\omega$ = angular velocity (radians/s), $S$ = sedimentation coefficient.

The sedimentation coefficient is measured in Svedbergs. One Svedberg unit is $10^{-13}$s, a cartilage proteoglycan aggregate has a sedimentation coefficient of about 109S whereas a cartilage proteoglycan monomer has a sedimentation coefficient of about 19S. The value of the sedimentation coefficient is determined by the molecular weight, shape and density of a particle in solution.

*Reagents:*

(A) 0.5 M Guanidine.HCl/1 mM EDTA (pH 5.8).

Dissolve 47.5 g guanidine.HCl and 0.372 g of EDTA in distilled water, adjust the pH to 5.8 with HCl and the final volume to 1 l.

(B) 4 M Guanidine.HCl/1 mM EDTA (pH 5.8).

Dissolve 380 g of guanidine.HCl and 0.372 g of EDTA in distilled water and adjust the pH to 5.8 with HCl and the final volume to 1 l.

*Method:*

(i)    Dissolve the proteoglycan preparation in reagent A to a final concentration of 200−300 $\mu$g uronic acid/ml. Dialyse this solution (or other proteoglycan) exhaustively against reagent A.

(ii)   Take the proteoglycan solution and centrifuge at 70 000−130 000 $g$ in a suitable analytical ultracentrifuge. In our laboratory we use an MSE Centriscan analytical ultracentrifuge with a six-place rotor using single-sector cells with a 2 cm light path length.

The centrifugation should be performed as follows: fill a single-sector cell with the proteoglycan test solution and fill a similar cell (reference cell) with the dialysate used for the proteoglycan solution. The meniscus of this tube should be level or slightly higher than the test cell. Centrifuge the cells at a speed such that the test sample completely sediments in about 90 min to 2 h. This speed must be determined empirically. One should aim in subsequent runs at this determined speed, to scan the sample about 6−10 times during the run. Also, scan the reference cell at several (known) times during the experiment.

Details of measurement of $r$ are determined in different ways for different ultracentrifuges. Consult the technical publications for precise instructions for measurement. After determination of the distance migrated ($r$), plot $\log_{10}r$ against time and determine the slope of the line. Using this value, the sedimentation coefficient may be determined thus:

$$S = \frac{2.303 \times 3600}{(2\pi \text{ r.p.m.})^2} \times \text{slope of line}$$

It is usual to convert this experimentally derived sedimentation coefficient ($S_{exp}$) to standard conditions which are 20°C, using water as solvent. Thus, the sedimentation coefficient under such standard conditions is referred to as $S_{20,W}$ and may be calculated using the $S_{exp}$ by the following relationship:

$$S_{20,W} = S_{exp} \frac{\eta_T}{\eta_{20}} \times \frac{(1-\nu\varrho)_{20,W}}{(1-\nu\varrho)_T}$$

where $\eta$ is the viscosity of the solvent at temperature $T$, $\eta_{20}$ is the viscosity of the solvent at temperature 20°C, $\nu$ is the partial specific volume of the solution, $\varrho$ is the density of the solvent (i.e. water at 20°C for the 20,W term, and the experimental solvent at temperature T is for the T term).

Under associative (0.5 M guanidine.HCl) conditions cartilage proteoglycans sediment as two separate species: (i) a fast sedimenting species corresponding to proteoglycan aggregates and (ii) a more slowly sedimenting species, corresponding to non-aggregating proteoglycans. *Figure 6* shows a diagrammatic representation of a sedimentation velocity

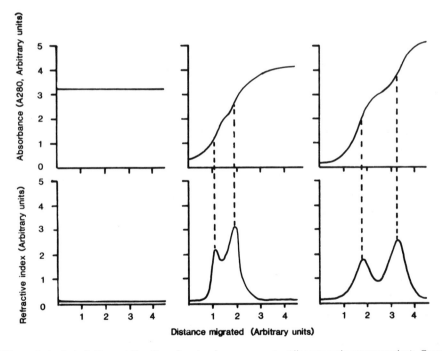

**Figure 6.** Analytical ultracentrifugation of proteoglycan aggregates (diagrammatic representation). Centrifugation of proteoglycans monitored by measurement of the absorbance at 280 nm ($A_{280}$; top panels) and change in refractive index (bottom panels) at (from left to right) time zero, at time 'x' min and at '1.5 x' min. The faster migrating species contains proteoglycan aggregates, the more slowly migrating species contain non-aggregating proteoglycans.

run of cartilage proteoglycan aggregates.

N.B. It is possible, although not common, to perform such sedimentation velocity procedures under dissociative (4 M guanidine.HCl) conditions. In this case, cartilage proteoglycans migrate as a single peak, corresponding to the cartilage proteoglycan monomer. To perform such runs, substitute reagent B for reagent A in all the steps described previously.

### 3.5 Analytical/semi-preparative centrifugation methods for proteoglycan examination

#### 3.5.1 *Rate zonal centrifugation in sucrose gradients* (17)

This technique relies upon the principle that larger macromolecules sediment at faster rates than corresponding smaller molecules. In this respect it is similar to sedimentation velocity centrifugation as described in Section 3.4.1. It is, however, far more akin to the moving band sedimentation velocity technique performed, in Vinograd cells, where one applies the sample as a narrow band and measures the passage of solutes through the cell as distinct zones. In this procedure, proteoglycans are centrifuged through a linear sucrose gradient from 10% to 50% (w/v) at high speed. Under these conditions, aggregating proteoglycans can easily be separated from non-aggregating proteoglycans which sediment far more slowly.

Heinegård (17) has further centrifuged the aggregating proteoglycans prepared by this method (after purification by dissociative CsCl isopycnic centrifugation) in 10−30% sucrose gradients to isolate two distinct aggregating populations from cartilage which are immunologically dissimilar and migrate separately on polyacrylamide−agarose gel electrophoresis (see Section 3.6).

*Reagents:*

(A) 50% Sucrose in 0.2 M $CaCl_2$/1 mM Tris-HCl (pH 7.5).

Dissolve 2.94 g of $CaCl_2$, 0.012 g of Tris and 50 g of sucrose in distilled water and adjust the pH to 7.5 with HCl and the final volume to 100 ml with distilled water.

(B) 0.2 M $CaCl_2$/1 mM Tris-HCl (pH 7.5).

Dissolve 2.94 g of $CaCl_2$ and 0.012 g of Tris in distilled water and adjust the pH to 7.5 with HCl and the final volume to 100 ml with distilled water.

*Method:*

(i)    Gradient preparation

1. By using reagents A and B prepare five solutions having sucrose concentrations of 50% (100% B, 0% A), 40% (80% B, 20% A), 30% (60% B, 40% A), 20% (40% B, 60% A) and 10% (20% B, 80% A).

2. Into the bottom of centrifuge tubes for a suitable swing out rotor (in this case a 6 × 14 ml MSE rotor) pipette 2.6 ml of the 50% sucrose solution into each. Carefully layer upon these 2.6 ml of the 40% solution, followed by 30%, 20% and 10%.

3. Allow these stepwise gradients to diffuse for about 2 h at 4°C to allow a linear continuous gradient from 50 to 10% sucrose to form.

(ii)   Sample preparation and application

1. Dissolve proteoglycan samples in reagent C to a concentration of about 5 mg/ml.

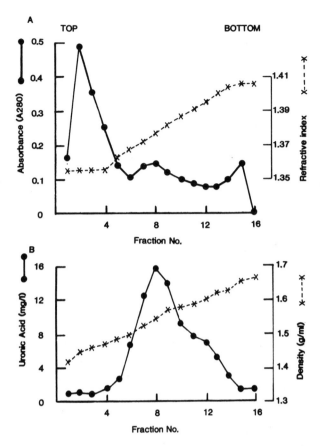

**Figure 7. A.** Rate zonal centrifugation of canine articular cartilage proteoglycans in sucrose gradients (this preparation has approximately 25% aggregating proteoglycans as assessed by gel chromatography). The peak near the top of the gradient contains the non-aggregating proteoglycans. The broader, more rapidly sediment-ing species corresponds to aggregating proteoglycans. **B.** Isopycnic ultracentrifugation in $Cs_2SO_4$. The pro-file illustrated indicates the equilibrium position of proteoglycan aggregates centrifuged in $Cs_2SO_4$ as described in Section 3.5.2.

    2. Carefully layer on top of the pre-formed gradients 1 ml of the proteoglycan sample.

(iii)   Centrifugation

    1. Centrifuge the samples at 200 000 $g$ for 55 min at 20°C.

    2. Recover fractions as described in Section 2.2.4. Sample results for centrifu-gation of proteoglycan aggregates are shown in *Figure 7A*.

N.B. When using different rotors, it may be necessary to modify the centrifugation conditions. This must be done empirically and depends on the proteoglycan being cen-trifuged and the dimensions of the rotor.

### 3.5.2 *Isopycnic ultracentrifugation in $Cs_2SO_4$*

The buoyant density of cartilage proteoglycans in CsCl is so high that it exceeds the maximum density that it is possible to obtain for CsCl in aqueous solution. As a result, the bulk of proteoglycans sediment at the bottom of such CsCl gradients. This is useful for bulk preparation of proteoglycans, but prevents examination of their polydispersity according to buoyant density. Several authors have reported that proteoglycans have a considerably reduced buoyant density in $Cs_2SO_4$. It would appear that such changes in buoyant density are due to reductions in the hydration of the proteoglycan. Reductions in buoyant density can be observed in CsCl gradients by increasing the guanidine.HCl concentration due to a cation effect of the $Cl^-$ species. A study of proteoglycan polydispersity using $Cs_2SO_4$ isopycnic centrifugation has been carried out by Bonnet *et al.* (18) from which this method has been taken.

*Method:*

(i)     Extract proteoglycans as outlined in Section 2.1.1.

(ii)    Dialyse to associative conditions as shown in Section 2.2.1.

(iii)   Prepare the required proteoglycan fraction (aggregate, monomer, etc.) as outlined in Section 2.2.

(iv)    Dialyse the proteoglycan preparations (0.5 mg/ml) against 0.5 M sodium acetate (pH 5.8).

(v)     Adjust the density of the proteoglycan solution to 1.53 g/ml by the addition of 0.635 g/ml solid $Cs_2SO_4$.

(vi)    Centrifuge at 100 000 *g* for 48 h at $10-20°C$.

(vii)   Fractionate the gradients as outlined in Section 2.2.4. Sample results are shown in *Figure 7B*.

N.B. As with all such procedures, conditions for centrifugation and starting density will vary from proteoglycan to proteoglycan. This set of procedures was derived for bovine nasal cartilage proteoglycan. The conditions for other proteoglycans will have to be determined empirically. The effect of guanidine.HCl on this separation is highly complex and if one wishes to perform such centrifugations under dissociative conditions, it is strongly advised that one reads the paper of Bonnet *et al.* (18) before commencing.

## 3.6 Gel electrophoresis

The efficiency of resolution of proteoglycan species by standard biochemical techniques is impaired due to the high degree of polydispersity of such molecules. The electrophoresis of proteoglycans on polyacrylamide-agarose gel matrices is capable, however, of resolving proteoglycan preparations into several (usually three or less) discrete bands (19). The separation is dependent upon factors such as molecular size and charge density, but exactly how the various bands are resolved is not clear.

The large molecular size of proteoglycans generally prevents their penetration into polyacrylamide gels. At concentrations below 2%, acrylamide will not form mechanically stable gels; however, even at this low concentration the pore size is too small to allow proteoglycans to enter the gel. To increase the pore size yet maintain rigidity it is necessary to include agarose at a concentration of about 0.5%. Empirically, it has been determined that optimum proteoglycan resolution is achieved at acrylamide concentrations of 1.2% (w/v) and agarose concentrations of 0.6% (w/v). This electrophoretic technique is useful for the qualitative examination of proteoglycans from different tissues and species, or alterations in proteoglycans produced by ageing or disease.

*Reagents:*

(A) 40 mM Tris-acetate/1 mM sodium sulphate (pH 6.8).

Dissolve 4.84 g of Tris and 0.142 g of $Na_2SO_4$ in distilled water and adjust the pH to 6.8 (at 4°C) with acetic acid and to a final volume of 1 l with distilled water.

(B) 6.4% (v/v) β-dimethylaminopropionitrile.

Mix 640 μl of β-dimethylaminopropionitrile with 9.36 ml of distilled water. This reagent must be prepared freshly.

(C) 3% (w/v) ammonium persulphate.

Dissolve 300 mg of ammonium persulphate in 10 ml of distilled water. This reagent must be prepared freshly.

(D) 30% (w/v) acrylamide (3% cross linking).

Dissolve 29.1 g of acrylamide and 0.9 g of bisacrylamide in distilled water to a final volume of 100 ml. This reagent is stable for 2 or 3 weeks if kept cool and protected from light.

(E) *n*-Butanol saturated with distilled water.


### 3.6.1 *Gel polymerization*

(i)   Wash gel plates (14 × 14 × 0.2 cm) thoroughly, dry and assemble casting apparatus.

(ii)  Mix 10 ml of reagent D with 18.47 ml distilled water and add 30 μl of TEMED and 1.5 ml of reagent C, mix rapidly and thoroughly then transfer to the gel plates to a depth of about 5 cm. Overlay this acrylamide plug with reagent E: after polymerization is complete, wash the gel surface with distilled water to remove *n*-butanol. Blot the gel plates dry with Whatman's No. 1 filter paper.

(iii) Add 0.24 g of agarose to 22.5 ml of reagent A and heat (with constant stirring) to boiling, then allow the agarose mixture to cool gradually (with constant stirring).

(iv)  Whilst the agarose cools, dissolve 0.456 g of acrylamide and 0.024 g of bisacrylamide in 9.7 ml of reagent A and add 4.8 ml of reagent B. Allow this mixture to warm by placing it next to the agarose mixture on the magnetic stirrer hot plate.

(v)   Whilst the agarose solution cools and the acrylamide solution warms, transfer the gel plates containing the polyacrylamide plug to an oven, preheated to 60°C.

(vi)  When the agarose solution has cooled to 52°C, add 3 ml of reagent C to the acrylamide solution, mix well and add to the agarose and continue mixing. Rapidly pour the agarose-acrylamide mixture into the gel plates.

(vii) Insert the well former and overlay with reagent E to exclude air. Maintain the gel at 4−8°C to ensure complete polymerization of the matrix.


### 3.6.2 *Electrophoresis*

(i)   Dissolve proteoglycan monomer preparations to a final concentration of 1 mg/ml in a 4-fold dilution of reagent A in distilled water containing 20% sucrose and 0.001% bromophenol blue.

(ii)    Place the gel slab (which has been washed with distilled water to remove all traces of *n*-butanol) into a suitable electrophoresis apparatus and remove the sample well former.

(iii)   Dilute reagent A 4-fold with distilled water and fill the electrophoresis reservoirs with this buffer.

(iv)    Apply $4-5$ $\mu$l proteoglycan samples under the buffer using a Hamilton syringe.

(v)     Electrophorese at 60 V (constant voltage) until the bromophenol blue tracking dye enters the gel, then increase the voltage to 120 V.

(vi)    Continue electrophoresis until the bromophenol blue tracking dye has migrated 3 cm.

### 3.6.3 *Detection*

Proteoglycans may be stained with a variety of dyes; however, the most common and possibly easiest method is using toluidine blue, and is outlined below.

(i)     Immerse the gel slab in a solution of 200 mg of toluidine blue in 100 ml of 0.1 M acetic acid for 20 min.

(ii)    Destain the gel in acetic acid (3% v/v) by soaking for 90 min.

(iii)   Destain until clear with several changes of distilled water. Proteoglycans appear as purple bands on a faint blue background under these conditions. *Figure 8A* shows an example of proteoglycans electrophoresed on such gels, stained with toluidine blue.

### 3.6.4 *Detection by fluorography* (20)

This technique may be used to visualize radiolabelled proteoglycans using this gel system, thus allowing the examination of newly-synthesized proteoglycans from tissue explant or cell cultures.

*Reagents:*

(A) Dissolve 0.246 g of sodium acetate in 200 ml of absolute ethanol.

(B) Dissolve 0.4 g of diphenyloxazole (PPO) in 100 ml of reagent A.

*Method:*

(i)     Fix the gel slab by immersion in 100 ml of reagent A for 16 h. This step may be performed on unstained gels or on gels that have been stained with toluidine blue as outlined in the previous section.

(ii)    Drain off reagent A and rinse the gel for 1 h in distilled water, with several changes.

(iii)   Dry the gel down on Whatman's No. 1 filter paper. This may be done using a commercially available gel drier, merely allowing the gel to dry in air will cause the gel to crack.

(iv)    Put the dried gel in contact with X-ray film, pre-flashed to an optical density of 0.2 at 530 nm (see ref. 21 for full details) and seal into light-tight boxes. All the procedures in this step *must* be performed in total darkness.

(v)     Develop the X-ray film at $-70°C$ for the required length of time. As a rule of thumb, [$^{35}$S]proteoglycans having 500 d.p.m. $^{35}$S require $4-5$ days development time.

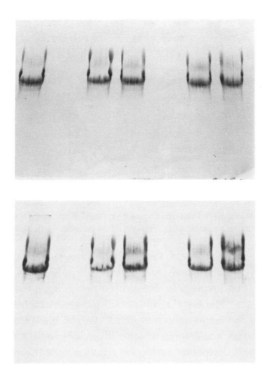

**Figure 8.** Electrophoresis of pig articular cartilage [$^{35}$S]-labelled proteoglycans on polyacrylamide-agarose gels. **A.** Stained with toluidine blue (as described in Section 3.6.3). **B.** The same gel visualized by fluorography (as described in Section 3.6.4).

(vi)  After development at $-70°$C, remove the X-ray film and develop as advised by the manufacturers (again, this step must be performed in total darkness). An example of visualization of $^{35}$S-labelled proteoglycans visualized by fluorography after electrophoresis is given in *Figure 8B.*

N.B. If there are no facilities for development at $-70°$C it is possible to carry out this step at $-20°$C. However, X-ray film is considerably less sensitive to the emitted photons at this temperature, therefore one either needs to increase exposure time (which generally leads to an increase in the background levels of the fluorogram) or to increase the radioactivity applied to the gel (which may not always be possible).

### 3.7 Depolymerization of glycosaminoglycans using glycanases

The depolymerization of proteoglycans using enzymes specific for the degradation of particular glycosaminoglycan species has become a popular method of analysis in recent years (22). The enzymes chondroitinase ABC (chondroitin ABC lyase) and chondroitinase AC (chondroitin AC lyase) have been most frequently used of all the glycosaminoglycanases. Chondroitin ABC will specifically cleave chondroitin 4-sulphate, chondroitin 6-sulphate, non-sulphated chondroitin and dermatan sulphate. Chondroitinase AC, however, will not degrade dermatan sulphate, although will degrade all of the other glycosaminoglycans mentioned previously. Thus one may differentiate between glyco-

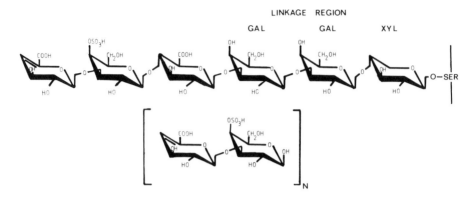

**Figure 9.** The structure of the delta-unsaturated chondroitin 4-sulphate 'stub' and the delta-unsaturated disaccharides produced by chondroitinase ABC digestion of chondroitin 4-sulphate described in Section 3.7.

saminoglycans containing dermatan sulphate and those lacking dermatan sulphate by digestion with AC lyase and subsequent digestion with ABC lyase.

The digestion reaction involves an elimination producing unsaturated disaccharides of chondroitin sulphate. The $C=C$ double bond is introduced between carbons 4 and 5 of the uronic acid residue as shown in *Figure 9*. The unsaturated disaccharide has a characteristic absorbance at 230 nm which allows one to easily monitor the eliminase reaction or alternatively detect the products of reaction after chromatographic fractionation, etc. The removal of glycosaminoglycan from proteoglycan by such techniques is not, however, complete and leaves a trisaccharide of chondroitin attached to the linkage region (*Figure 9*). The 'core-protein preparations' produced by such enzymic depolymerization can be easily examined by standard SDS-PAGE or gel chromatography. This therefore provides a technique by which relative proteoglycan core protein sizes may be compared (23). It must be remembered, however, that such preparations are still highly glycosylated and hence will be considerably larger than the actual core protein size.

The introduction of a $C=C$ double bond in the terminal glucuronic acid residue of the trisaccharide still attached to the protein core via the linkage region renders this residue highly antigenic. Antibodies may be raised against such unsaturated 'stubs' of either chondroitin 4- or chondroitin 6-sulphate (24). After depolymerization of dermatan sulphate, however, it is indistinguishable from chondroitin sulphate, since the elimination reaction removes the asymmetry from the carbon atom at C-5 of the hexuronic acid residue.

*Reagents*

(A) 0.1 M Tris-HCl buffer (pH 8.0).

Dissolve 1.21 g of Tris, 0.024 g of sodium acetate, 0.01 g of bovine serum albumin, 0.372 g of EDTA and 0.125 g of NEM in distilled water. Adjust the pH to 8.0 with HCl and the volume to 99 ml. Just before use add PMSF (17 mg in 0.5 ml methanol) and pepstatin (16 mg in 0.5 ml methanol).

(B) 1 M Tris-HCl buffer (pH 7.4).

Prepared exactly as for reagent A but the pH adjusted to 7.4 rather than 8.0.

*Method:*

(i)     *For chondroitinase ABC.* Dissolve proteoglycan ($\sim$ 200 $\mu$g as uronic acid) in 1 ml of reagent A. Add to this solution chondroitinase ABC (0.05 units) and incubate at 37°C for 40 min. After incubation, the reaction may be terminated by the addition of an equal volume of 8 M guanidine.HCl in reagent A.

(ii)    *For chondroitinase AC.* Dissolve proteoglycan ($\sim$ 200 $\mu$g as uronic acid) in 1 ml of reagent B. Add to this solution chondroitinase AC (0.05 units) and incubate at 37°C for 40 min. The reaction may be terminated as outlined above.

As mentioned previously, the generation of $\Delta$di 4-5 unsaturated disaccharides can be monitored by the measurement of the absorbance of the solution at 230 nm. Thus one may increase or decrease either incubation time or enzyme or concentration as may be convenient and merely monitor the reaction until it reaches completion, since different proteoglycans may be depolymerized at widely varying rates.

Other enzymes are commercially available for the depolymerization of other glycosaminoglycans particularly keratanase (endo $\beta$-D-galactosidase) and heparatinase (21,25). Keratanase may be used exactly as described for chondroitinase ABC although the nature of the cleavage of keratanase and the structure of the limit digest product of skeletal keratan sulphate is not certain. Other enzymes active against glycans are becoming more freely available and one should carefully observe manufacturers' recommendations before use on proteoglycan molecules.

## 4. PROTEOGLYCAN CONSTITUENT ANALYTICAL TECHNIQUES

### 4.1 Uronic acid (carbazole) assay

This assay is essentially the modification of Bitter and Muir (26) of an older method of Dische (27) which involves the reaction of carbazole with the unstable acid hydrolysed dehydrated derivatives of hexuronic acids. This technique allows the estimation of glucuronic acids, found in chondroitin sulphates and hyaluronic acid, and also iduronic acids, found in dermatan sulphate and heparin. Heparin (and heparan sulphate), however, gives anomalously high results using this method probably due to the production of free amino residues liberated from the *N*-sulphate groups following acid hydrolysis.

### 4.1.1 *High sensitivity manual method*

*Reagents:*

(A) Borate sulphuric acid reagent. Dissolve 2.39 g of sodium tetraborate decahydrate in 250 ml of concentrated $H_2SO_4$. It is difficult to dissolve borate in sulphuric acid and warming produces only slow increases in the rate of dissolution (and is potentially hazardous), therefore it is desirable to prepare this reagent 1 day before use and leave to dissolve overnight. High-grade sulphuric acid should be used for this assay since the presence of heavy metal ions causes anomalous colour development.

(B) Carbazole. Dissolve 125 mg of carbazole (Analar grade) in 100 ml of absolute ethanol. Store in the dark, refrigerated and well stoppered to exclude air. Both reagents should be of analytical grade or better.

(C) Glucuronolactone standards. Glucuronolactone (1 mg/ml) dissolved in water saturated with benzoic acid.

*Method:*

(i) Add 1500 $\mu$l of reagent A to screw-capped tubes and cool in ice. Teflon tape may be used to line the caps of these tubes to prevent corrosion.

(ii) Carefully layer standard or test solutions (250 $\mu$l, containing <20 $\mu$g uronic acid/ml) onto the surface of the borate-sulphuric acid mixture and allow to diffuse for 10 min.

(iii) Gently mix the tubes, keeping cool to prevent excessive increases in temperature which may cause caramelization of the carbohydrate. When thoroughly mixed, briefly vortex mix to ensure homogeneity.

(iv) Heat the tubes in a boiling water bath for 10 min, then cool in an ice bath.

(v) Add cooled carbazole (reagent B, 50 $\mu$l) to the mixture, vortex mix and return to the boiling water bath for 15 min.

(vi) Cool in an ice-water bath, then read the absorbance at 525 nm.

Under these conditions, neutral sugars at similar concentrations produce about 10% interference; glucuronolactone (20 $\mu$g/ml) has an $A_{525}^{1\,cm}$ of 1.0 in this assay.

### 4.1.2 *Lower sensitivity manual method*

*Reagents:*

As used in the previous assay.

*Method:*

(i) Mix 1500 $\mu$l of reagent A with the sample of standards (250 $\mu$l) and 50 $\mu$l of reagent B. During mixing, the tubes must be cooled in an acetone salt-ice bath (at about $-20°C$) to prevent rapid increases in temperature. Providing initial mixing is quite gentle, caramelization of the sample is avoided.

(ii) Heat the tubes in a boiling water bath for 25 min.

(iii) Cool the tubes in an ice bath.

(iv) When cool, read the absorbance at 525 nm.

This method is considerably less sensitive than the previous method but has the advantage of only one heating and cooling step and hence is quicker than the high-sensitivity assay.

### 4.1.3 *Automated uronic acid assay*

The method is essentially the modification of the assay of Bitter and Muir (26) automated by the procedure of Heinegård (28).

Automated procedures are given for the assays of hexuronic acid, hexosamine and neutral sugar. The automated assays described below use Technicon autoanalysis equipment although other equipment may be used. Such systems allow the rapid and accurate determination of a number of chemical constituents of proteoglycan. In most cases, the board diagrams are given for the Technicon AAI using a peristaltic multichannel proportionating pump. The chromogen is detected using a continuous flow colorimeter, which may be equipped with a curve regenerator. The curve regenerator is an electronic device which detects the signal from the colorimeter and determines the final peak height from the rate of change of signal. This allows more samples to be processed per unit time since one does not have to let the chromogen decay to base levels

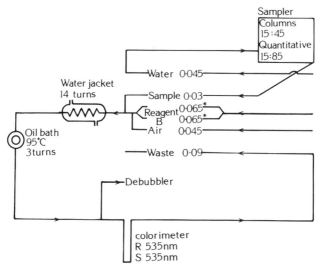

**Figure 10.** Board diagram for the automated analysis of uronic acid using the Technicon AAI system. The reference filter and sample filter are both 535 nm (R535 nm and S535 nm).

before one may assay the next sample. Board diagrams are occasionally given for the Technicon AAII system, which is an update of the AAI system. In essence the principles behind this system are identical to the AAI, but the AAII has the following advantages:

(i)     it requires less sample;
(ii)    it uses less reagent;
(iii)   it is considerably more sensitive than the AAI.

In all the board diagrams, the figures on the tubing refer to the tube internal diameter in inches. Those tubes marked with an asterisk are acidflex tubes, all other tubing used is solvaflex. The figures on the samples refer to the sample and wash times for the sampler turntable. The first figure always refers to the sample time (in seconds) and the second to the wash time (in seconds). The sets of figures given for columns refer to suitable settings for analyses of chromatographic column eluant fractions, whereas the settings marked quantitative should be used if one wishes to make accurate quantitative determination of a particular compound.

*Reagents:*

(A) Sodium tetraborate decahydrate (23.9 g) added to 2.5 l concentrated sulphuric acid (exactly as described for the manual methods).

(B) Carbazole (48 mg) added to 100 ml reagent A, stirred for about 20 min before use.

(C) Glucuronolactone (1 mg/ml) as standard stock solution (as used for manual assay).

*Method:*

For Technicon AAI. See *Figure 10*.
For Technicon AAII. See *Figure 11*.

N.B. Both the AAI and AAII methods are capable of assaying quantities of uronic acid

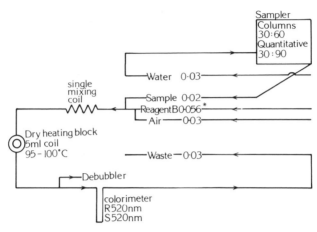

**Figure 11.** Board diagram for the automated analysis of uronic acid using the Technicon AAII system. The reference filter and sample filter (R520 nm and S520 nm) are both 520 nm. (As adapted by Dr M.T.Bayliss.)

in the range $1-25$ $\mu g/ml$. Unlike the manual assays, it is not possible to assay samples containing CsCl (as might be obtained from density gradient centrifugation of proteoglycans) due to the liberation of chlorine gas which severely disrupts the bubble pattern and prevents even flow through the colorimeter.

## 4.2 Glycosaminoglycan estimation by dye binding assays

These types of GAG assay have become increasingly popular, probably reflecting the relative ease and safety of these assays compared with the carbazole-sulphuric acid (uronic acid) assay. These types of assay can be divided into two categories: the precipitation type assay and the metachromatic type of assay. Both types of assay depend upon the electrostatic interaction of GAG and dye; in one case precipitation of the GAG-dye complex and subsequent dissociation of dye from GAG in stoichiometric amounts and its colorimetric estimation forms the basis of the assay. The other type of assay relies on the alteration of the absorbance spectrum of the dye by the complex formed with GAG. Measurement of the change in absorption at a single wavelength provides the basis of the metachromatic type of assay (*Figure 12*).

### 4.2.1 Precipitation assay

This assay is essentially that described by Whiteman (29) and relies upon the precipitation of Alcian Blue-GAG complexes and the dissociation of this dye from the complex by the use of suitable surfactants. The inclusion of magnesium chloride is important in this assay since at low concentrations of this salt all GAG species will bind to Alcian Blue. Increasing the concentration of $MgCl_2$ will prevent the binding of certain GAGs to Alcian Blue by a critical electrolyte concentration (CEC) technique, which thus allows the assay (if required) of particular GAG species (see ref. 30). The assay outlined has a $MgCl_2$ concentration such that all GAGs will bind to Alcian Blue.

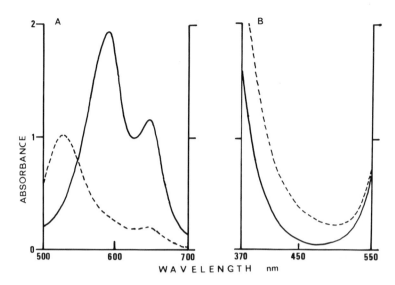

**Figure 12.** The absorbance spectra of metachromatic dyes. **A.** 1,9-Dimethylmethylene blue in the absence (—) and presence (- - -) of chondroitin sulphate. **B.** Alcian Blue in the absence (—) and presence (- - -) of chondroitin sulphate.

*Reagents:*

(A) Dissolve 50 mg of Alcian Blue in 100 ml of 50 mM sodium acetate/50 mM $MgCl_2$ (pH 5.8).

(B) Dissolve 40 g of Manoxol 1B in 50 mM sodium acetate (pH 5.8) (final volume 100 ml) *or* 10 g of sodium dodecyl sulphate in 50 mM sodium acetate (pH 5.8) (final volume 100 ml).

(C) Dissolve chondroitin 4-sulphate in distilled water to a concentration of 1 mg/ml as a stock standard solution.

*Method:*

(i)  Add solutions of standards or test samples (0.5 ml) to 4 ml of freshly prepared reagent A. Allow these mixtures to stand for 2 h at room temperature, with occasional vortex stirring.

(ii)  Centrifuge the tubes at 2000 *g* for 15 min.

(iii)  Discard the supernatant and resuspend the precipitate in 4 ml of absolute ethanol.

(iv)  Centrifuge the tubes at 2000 *g* for 15 min.

(v)  Discard the supernatant and dissociate the dye-GAG precipitate with 4 ml of reagent B.

(vi)  Vortex-mix the tubes and read the absorbance at 620 nm.

N.B. It may not be possible to obtain Manoxol 1B (sodium sulphosuccinic acid butyl ester); however, the related compound, Manoxol OT (the dioctyl ester of the above salt) is *not* suitable for the dissociation of the dye-GAG complex. As an alternative, sodium dodecyl sulphate (10% w/v) may be used. Under these conditions, chondroitin sulphate at a concentration of 25 $\mu$g/ml gives an $A_{620}^{1\,cm}$ of 0.6.

### 4.2.2 *Metachromatic assays*

*Using Alcian Blue by the method of Gold (31)*

*Reagents:*

(A) Dissolve Alcian Blue in 0.5 M sodium acetate to a concentration of 1.4 mg/ml. This reagent must be freshly prepared.

(B) Chondroitin 4-sulphate (1 mg/ml) in distilled water as a standard stock solution.

*Method:*

(i)   Mix standard or test solutions (100 $\mu$l) with 1.2 ml of reagent A. Leave to stand for 10 min.

(ii)  Read the absorption at 480 nm.

This assay is stable at low salt concentrations but less reliable at high salt concentrations. The assay is linear from 0 to 100 $\mu$g/ml chondroitin 4-sulphate, 50 $\mu$g/ml giving an $A_{480}^{1 cm}$ of about 0.2 This procedure is relatively insensitive; however, it is useful for initial determination of GAG content in unknown samples due to its wide range and rapidity. One may then choose to adopt a more sensitive assay for subsequent estimation of GAG content.

*Using 1,9-dimethylmethylene blue by the method of Farndale et al. (32)*

*Reagents:*

(A) Dissolve 16 mg of 1,9-dimethylmethylene blue in 5 ml of absolute ethanol. To this solution, add 2.0 g of sodium formate and 2 ml of formic acid (90% w/v) and distilled water to a final volume of 1 l (pH 3.5).

(B) Chondroitin sulphate (1 mg/ml) as standard stock solution.

*Method:*

(i)   Mix reagent A (2.5 ml) with standard or test solutions (250 $\mu$l) containing < 40 $\mu$g/ml chondroitin sulphate (or equivalent).

(ii)  Vortex-mix and immediately read the absorbance at 535 nm.

The dye solution in this assay is stable for several weeks if maintained refrigerated and in the dark. It is possible, using this assay, to estimate GAG concentrations in high salt (up to 2 M guanidine.HCl for example). However in this case all standard solutions should be adjusted to the salt concentration of the tests.

The metachromatic assays are suitable for sulphated glycosaminoglycans but not non-sulphated glycosaminoglycans. Hyaluronic acid, for example, gives only one-fifth of the absorbance of equivalent amounts of chondroitin sulphate.

Care must be taken to ensure that dye-GAG complex does not precipitate during the assay. At high GAG concentrations, complex precipitation may occur within 10 min, hence one should not attempt to assay large numbers of samples at a time. It is prudent to assay in batches with numbers which can comfortably be read in 10 min.

### 4.3 **The Elson–Morgan assay for hexosamine** (33)

### 4.3.1 *Manual assay*

*Reagents:*

(A) Dissolve 1 ml of redistilled acetylacetone in 50 ml of 0.5 M sodium carbonate.

(B) Dissolve 0.8 g of 4-(*N*,*N*-dimethylamino)benzaldehyde in 30 ml of absolute ethanol and mix with 30 ml of concentrated HCl.

(C) Dissolve 10 mg of 2-amino-2-deoxy-D-glucose-HCl (glucosamine-HCl) in distilled water as a standard.

*Method:*

(i) Hydrolyse proteoglycan samples in 8 M HCl for 4 h at 95°C. Adjust to pH 10 and to a known volume by addition of 4 M NaOH and distilled water respectively.

(ii) Mix the proteoglycan hydrolysate (or standard) solution (250 $\mu$l, containing <80 $\mu$l hexosamine) with 250 $\mu$l of reagent A and adjust the solution to a final volume of 600 $\mu$l with distilled water.

(iii) Stopper the tubes and heat at 100°C for 20 min, then allow to cool to room temperature.

(iv) Add 1 ml of absolute ethanol, taking care to wash down all droplets of condensation into the bottom of the tube.

(v) Add 250 $\mu$l of reagent B and dilute to a final volume of 2.4 ml with absolute ethanol.

(vi) Heat the tubes at 65°C for 10 min, cool to room temperature and determine the absorbance at 530 nm.

Under these conditions, 50 $\mu$g of 2-amino-2-deoxy-D-glucose-HCl gives an $A_{530}^{1\ cm}$ of 0.7.

### 4.3.2 *Automated Elson–Morgan assay (for Technicon AII)*

*Reagents:*

(A) Dissolve 3.5 g of acetylacetone in a solution of trisodium phosphate (11.4 g in 100 ml).

(B) Dissolve 3.2 g of 4-(*N*,*N*-dimethylamino)benzaldehyde in 30 ml of absolute ethanol and mix with 30 ml concentrated HCl. Just before use, dilute to 180 ml with absolute ethanol.

(C) Standard hexosamine. As used in manual assay.

For the assay method for Technicon AAI see *Figure 13*.

## 4.4 **Anthrone reaction for neutral sugars** (34)

This assay may be used to monitor column effluents for the presence of keratan sulphate in conjunction with the carbazole assay for uronic acids as mentioned in Section 1.1.2. The presence of keratan sulphate in column effluents produces an increase in the ratio of neutral sugar compared with uronic acid.

*Reagent:*

(A) Add 500 ml of concentrated $H_2SO_4$ to 200 ml of distilled water. Dissolve 0.28 g of anthrone in a portion of the diluted $H_2SO_4$ (100 ml). This reagent must be prepared freshly.

(B) D-galactose. Dissolve 0.2 g of galactose in 100 ml of distilled wter. This serves as a stock standard solution and should be stored at $-20°C$.

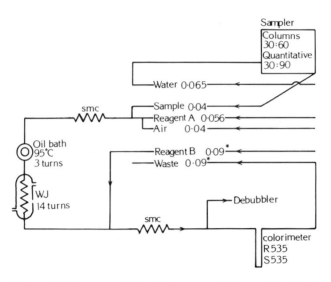

**Figure 13.** Board diagram for the automated Elson–Morgan assay for hexosamine using the Technicon AAI system. Both the reference and sample filters are 535 nm. smc: single mixing coil. wj: water jacket.

### 4.4.1 *Manual assay*

*Method:*

(i)     Add 5 ml of reagent A to suitably sized stoppered boiling tubes and keep cool by immersion in an ice/salt-water cooling mixture.

(ii)    Layer onto the surface of the cooled reagent A, sample or standard solutions (1.0 ml) containing <100 μg of neutral sugar.

(iii)   After allowing the sample and reagent A to diffuse together (10–15 min), mix well and heat for 10 min at 100°C.

(iv)    Cool the tubes in an ice-water bath and read the absorbance at 620 nm.

### 4.4.2 *Automated anthrone assay for neutral sugar*

The board diagram for Technicon AAII assay is identical to that shown for uronic acid (*Figure 11*) except that anthrone reagent (reagent A used for the manual anthrone assay) substitutes for the carbazole reagent used in that assay, and that the colorimeter should be set for 630 nm reference and 630 nm sample (as opposed to 520 nm in the uronic acid assay).

### 4.5 **Assay for sulphate groups** (35)

*Reagents:*

(A) Dissolve 4 g of trichloroacetic acid in distilled water (final volume 100 ml).

(B) Gelatin/$BaCl_2$ reagent. Dissolve 2.0 g of gelatin in 400 ml of hot (60–70°C) distilled deionized water. Cool and allow to stand for at least 6 h at 4°C. Dissolve 20 g of $BaCl_2$ in this gelatinous solution and allow to stand for 2–3 h before use. Kept cool (4°C), the reagent is stable for about 1 week.

(C) Dissolve 1.813 g of $K_2SO_4$ in distilled water (final volume 100 ml). This solution is used as standard and contains 10 mg $SO_4^{2-}$ per ml.

*Method:*

(i)    *Hydrolysis of samples*

    1. Adjust the proteoglycan solution to 60% (w/v) formic acid, by the addition of 90% (w/v) formic acid.

    2. Hydrolyse the sample (in tightly sealed tubes) for 8 h at 100°C.

    3. Dry the sample over KOH pellets, in a desiccator maintained under vaccum.

    4. Resuspend the dried hydrolysate in distilled water (0.5 ml).

(ii)    *Assay method*

    1. Mix the hydrolysate (or standard solution; 0.2 ml) with reagent A (3.8 ml) and reagent B (1 ml).

    2. Allow to stand for 15 min, then read the opacity of the solution at 500 nm.

    3. The values obtained may be compared with a standard curve using $K_2SO_4$ (reagent C). The assay is linear between 20 and 200 $\mu$g of $SO_4^{2-}$.

N.B. All glassware should be cleaned with concentrated $HNO_3$ before use. Since phosphate interferes with this assay one must be particularly careful to avoid buffers containing this ion during the preparation and isolation of samples for sulphate estimation.

## 4.6 Assay for N-sulphate groups (36)

Heparin and heparan sulphate are unique compared with the other glycosaminoglycan species in that they alone contain *N*-sulphated hexosamines. Nitrous acid reacts with *N*-sulphated-D-glucosamine in heparin (and heparan sulphate) preparations, cleaving the chain after this residue to leave a 2,5-anhydromannose residue at the reducing terminus of oligosaccharides produced by such cleavage. The mechanism of this reaction is shown in *Figure 14*.

One may therefore estimate *N*-sulphate by an indirect method, that is, by the determination of 2,5-anhydromannose following hydrolysis by nitrous acid. Although *N*-acetylated-D-glucosamine may be similarly deaminated to yield anhydromannose residues, at dilute acid concentrations *N*-acetyl groups are quite stable in comparison with

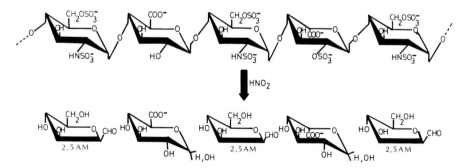

**Figure 14.** Cleavage of heparin or heparan sulphate by nitrous acid, showing the stoichiometric conversion of *N*-sulphated residues to free 2,5-anhydromannitol which may be determined as outlined in Section 4.6. 2,5 AM: 2,5-anhydromannitol.

*N*-sulphate, which thus allows one to differentiate between these two *N*-substituted derivatives of D-glucosamine.

*Reagents:*

(A) Mix 1 ml of butylnitrite with 19 ml of absolute ethanol. This reagent must be prepared freshly.

(B) Mix 3.436 ml of concentrated HCl with 16.564 ml of distilled water.

(C) Dissolve 1.25 g of ammonium sulphamate in distilled water (final volume, 10 ml).

(D) Mix 5 ml of concentrated HCl with 95 ml of distilled water.

(E) Dissolve 5 mg of indole in 10 ml of absolute ethanol.

*Method:*

(i)     Mix the glycosaminoglycan test or standard solution ($<50$ $\mu$g/ml uronic acid; 200 $\mu$l) with 200 $\mu$l of reagent A and 200 $\mu$l of reagent B and 200 $\mu$l of reagent B. Incubate at room temperature for 1 h.

(ii)    Add 200 $\mu$l of reagent C and incubate for 1 h at room temperature with occasional stirring.

(iii)   Add 1.0 ml of reagent D and 200 $\mu$l of reagent E and incubate for 5 min at 100°C.

(iv)    Cool the tubes and add 1.2 ml of absolute ethanol. Read the absorption at 520 nm.

## 4.7 Liquid phase competition radioimmunoassay for proteoglycan substructures (37,38)

Until the development of specific antibodies directed against the hyaluronic acid-binding region and link proteins of proteoglycan aggregates, there was no assay available for those components. The use of immunoassays has been invaluable in examinations of proteoglycan biosynthesis and degradation and more recently has been used to differentiate different subspecies in preparations of proteoglycan.

*Reagents:*

(A) Dissolve 0.5 g of sodium deoxycholate, 0.1 g of Nonidet NP-40, 0.1 g of bovine serum albumin and 0.03 g of sodium azide, in phosphate-buffered saline [0.15 M NaCl buffered with 0.01 M sodium phosphate (pH 7.4)] to a final volume of 100 ml.

(B) Suspend 500 mg of heat-killed formaldehyde-treated *S. aureus* in 5 ml of reagent A. Spin down (20 000 $g$, 4 min), discard the supernatant, resuspend and sonicate in 5 ml of reagent A. *S. aureus* is available commercially in a freeze-dried form from Miles Laboratories (Stoke Poges, UK).

(C) [125]I-labelled antigens (either link protein or hyaluronic acid-binding region). The preparation of these reagents is complicated, and is described in full detail in Section 3.3.

(D) Antiserum (or monoclonal antibodies) directed against the respective antigen [details of antiserum production are given in various papers, notably by Caterson *et al.* (37) and Ratcliffe and Hardingham (38)].

N.B. Reagents B, C and D are all dissolved or suspended in reagent A.

### 4.7.1 *Titration of antisera*

(i)     Incubate a dilution of reagent C in reagent A (100 $\mu$l, containing $\sim$3 ng protein

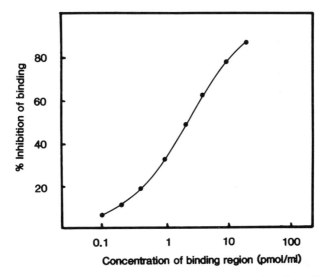

**Figure 15.** An antiserum titration curve for anti-(hyaluronic acid-binding region). From this one may determine the antibody dilution capable of binding approximately 50% of the maximum binding of antigen to antibody. This dilution of antiserum should then be used for preparation of a standard curve (*Figure 16*) and estimation of unknowns.

and 10 000 – 20 000 c.p.m.) with dilutions of reagent D in reagent A (from 1/4 to 1/2048; 100 $\mu$l) for 90 min at room temperature.

(ii)   Add reagent B (100 $\mu$l), mix, and incubate for 15 min at room temperature.

(iii)  Add reagent A (1.0 ml), mix, and centrifuge (12 000 $g$, 4 min) to pellet $^{125}$I antigen-antibody complexes.

(iv)   Carefully decant off and retain the supernatant. Wash the pellet once with reagent A (1.0 ml).

(v)    Recentrifuge the suspension (12 000 $g$, 4 min), decant off the supernatant and combine with the supernatant obtained from step (iv).

(vi)   Determine the radioactivity in the supernatant (non-bound) and pellet (bound) using a suitable gamma counter.

(vii)  Perform steps (i) – (vi) in the absence of antibody to determine the background levels of $^{125}$I to substract from experimental results.

(viii) Express the results as percentage of $^{125}$I-labelled antigen bound versus the dilution of antiserum. A typical curve is shown in *Figure 15*. A dilution of the antiserum which gives a value of approximately 50% of maximal binding should then be used for subsequent preparation of standard curves and determination of unknowns.

### 4.7.2 *Preparation of standard curve*

(i)    Mix $^{125}$I-labelled antigen (10 000 – 20 000 c.p.m., 50 $\mu$l) with a known amount (10 – 1000 ng) of unlabelled antigen (50 $\mu$l). Heat for 20 min at 80°C.

(ii)   Add an antiserum dilution (500 $\mu$l) which was determined in step (vii) of Section 4.7.1. Mix.

(iii)  Incubate at room temperature for 2 h.

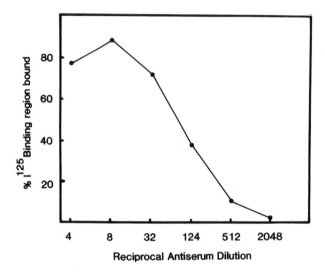

**Figure 16.** The inhibition of binding of $^{125}$I-HABR to anti-(HABR) by known amounts of unlabelled HABR using the antiserum dilution determined in *Figure 15*. From this standard curve one may estimate the amount of HABR in unknown samples by the determination of the inhibition of binding produced by such samples. Data for *Figures 15* and *16* courtesy of Dr A.Ratcliffe.

(iv)    Add reagent B (30 $\mu$l), centrifuge (12 000 $g$, 4 min), decant and retain the supernatant.

(v)    Wash the pellet with reagent A (1.0 ml), centrifuge (12 000 $g$, 4 min), decant off the supernatant and combine with the supernatant obtained from step (iv).

(vi)    Determine the radioactivity in the supernatant (non-bound) and pellet (bound) using a suitable gamma counter.

(vii)    Express the results as percentage inhibition of binding of $^{125}$I-labelled antigen versus concentration of non-radioactive antigen. A typical curve is shown in *Figure 16*.

### 4.7.3 *Estimation of unknowns*

(i)    Prepare a range of dilutions in reagent A of the unknown samples to be assayed (50 $\mu$l) and mix with $^{125}$I-labelled antigen (100 $\mu$l). Repeat steps (ii)−(vi) in Section 4.7.2.

(ii)    By determination of the percentage inhibition (between about 20 and 60%: the most accurate portion of the standard inhibition curve) of binding produced by the test (unknown) samples, one may calculate from the standard curve the amount of antigen in each dilution.

N.B. It is possible to assay for epitopes of the proteoglycan which are normally sequestered in the proteoglycan ternary complex structure, by unfolding and dissociating the protein structures by heating in SDS (0.025% w/v). The details of this modified assay are given in Ratcliffe and Hardingham (38).

If the antibodies raised against particular proteoglycan substructures do not bind to

Staph A, a second antibody step must be included. This second antibody must bind to Staph A, in addition to the anti-proteoglycan antibody. A dilution of 1:100 (20 $\mu$l) is generally sufficient for this purpose.

## 5. ACKNOWLEDGEMENTS

I thank Miss Véronique Hazan for the excellent preparation of this manuscript and all my colleagues in the Biochemistry Division of the Kennedy Institute. Particular thanks for help and advice must be extended to Dr Helen Muir, Dr Tim Hardingham, Dr Mike Bayliss, Dr Tony Ratcliffe and Mr Dave Dunham, whose assistance was invaluable.

## 6. REFERENCES

1. Heinegård,D., Björne-Persson,A., Cöster,L., Franzén,A., Gardell,S., Malmström,A., Paulsson,M. and Vogel,K. (1985) *Biochem. J.*, **230**, 181.
2. Rodén,L. (1980) In *Biochemistry of Glycoproteins and Proteoglycans*, Lennarz,E. (ed.), Plenum, New York, p. 267.
3. Heinegård,D. and Paulsson,M. (1984) In *Extracellular Matrix Biochemistry*, Piez,K.A. and Reddi,A.H. (eds), Elsevier, New York, Amsterdam, Oxford, p. 277.
4. Hardingham,T.E. (1986) In *Connective Tissue in Normal and Pathological States*, Kuhn,K. and Krieg,T. (eds), Karger, Basel, p. 143.
5. Comper,W.D. and Laurent,T.C. (1978) *Phys. Rev.*, **58**, 255.
6. Bayliss,M.T., Venn,M., Maroudas,A. and Ali,S.Y. (1983) *Biochem. J.*, **209**, 387.
7. Byers,S., Kuettner,K.E. and Kimura,J.H. (1985) In *Transactions of the 31st Annual Meeting of the American Orthopaedic Research Society*, Las Vegas, Nevada, p. 50.
8. Christner,J.E., Baker,J.R. and Caterson,B. (1983) *J. Biol. Chem.*, **258**, 14335.
9. Franek,M.D. and Dunstone,J.R. (1966) *Biochim. Biophys. Acta*, **127**, 213.
10. Sajdera,S.W. and Hascall,V.C. (1969) *J. Biol. Chem.*, **244**, 77.
11. Antonopoulos,C.A., Axelsson,I., Heinegård,D. and Gardell,S. (1974) *Biochim. Biophys. Acta*, **338**, 108.
12. Hardingham,T.E. (1979) *Biochem. J.*, **177**, 237.
13. Kimura,J.H., Hardingham,T.E., Hascall,V.C. and Solursh,M. (1979) *J. Biol. Chem.*, **254**, 2600.
14. Hascall,V.C. and Heinegård,D. (1974) *J. Biol. Chem.*, **249**, 4242.
15. Heinegård,D. and Hascall,V.C. (1974) *J. Biol. Chem.*, **249**, 4250.
16. Bonnet,F., Dunham,D.G. and Hardingham,T.E. (1985) *Biochem. J.*, **228**, 77.
17. Heinegård,D., Wieslander,J., Sheehan,J., Paulsson,M. and Sommarin,Y. (1985) *Biochem. J.*, **225**, 95.
18. Bonnet,F., Périn,J.-P. and Jollès,P. (1980) *Biochim. Biophys. Acta*, **623**, 57.
19. McDevitt,C.A. and Muir,H. (1971) *Anal. Biochem.*, **44**, 612.
20. Carney,S.L., Bayliss,M.T., Collier,J.M. and Muir,H. (1986) *Anal. Biochem.*, **156**, 38.
21. Laskey,R.A. and Mills,A.D. (1975) *Eur. J. Biochem.*, **56**, 335.
22. Saito,H., Yamagata,T. and Suzuki,S. (1968) *J. Biol. Chem.*, **243**, 1536.
23. Oike,Y., Kimata,K., Shinomura,T., Nakazawa,K. and Suzuki,S. (1980) *Biochem. J.*, **191**, 193.
24. Caterson,B., Christner,J.E., Baker,J.R. and Couchman,J.R. (1985) *Fed. Proc.*, **44**, 386.
25. Kato,M., Oike,Y., Suzuki,S. and Kimata,K. (1985) *Anal. Biochem.*, **148**, 479.
26. Bitter,T. and Muir,H.M. (1962) *Anal. Biochem.*, **4**, 330.
27. Dische,Z. (1947) *J. Biol. Chem.*, **167**, 189.
28. Heinegård,D. (1973) *Chem. Scr.*, **4**, 199.
29. Whiteman,P. (1973) *Biochem. J.*, **131**, 343.
30. Scott,J.E. and Dorling,J. (1965) *Histochemie*, **5**, 221.
31. Gold,E.W. (1979) *Anal. Biochem.*, **99**, 183.
32. Farndale,R.W., Sayers,C.A. and Barrett,A.J. (1982) *Conn. Tis. Res.*, **9**, 247.
33. Elson,L.A. and Morgan,W.T.J. (1933) *Biochem. J.*, **23**, 1824.
34. Trevelyan,W.E. and Harrison,J.S. (1952) *Biochem. J.*, **51**, 298.
35. Dodgson,K.S. (1961) *Biochem. J.*, **78**, 312.
36. Dische,Z. and Borenfreund,E. (1950) *J. Biol. Chem.*, **184**, 517.
37. Caterson,B., Baker,J.R., Levitt,D. and Paslay,J.W. (1979) *J. Biol. Chem.*, **231**, 9369.
38. Ratcliffe,A. and Hardingham,T. (1983) *Biochem. J.*, **213**, 371.

CHAPTER 5

# Glycoproteins

JEAN MONTREUIL, STÉPHANE BOUQUELET, HENRI DEBRAY,
BERNARD FOURNET, GENEVIÈVE SPIK and GÉRARD STRECKER

## 1. INTRODUCTION

Glycoproteins (for recent reviews, see ref. 1−6), which together with glycolipids constitute the class of glycoconjugates, result from the covalent association of carbohydrate moieties (glycans) with proteins. Glycosylation of proteins represents one of the most important post-translational events because of the universality of the phenomenon. Most proteins are glycosylated, the glycoproteins being widely distributed in animals, plants, microorganisms and viruses. It has been established that glycans perform important biological roles including:

(i)     the protection of the peptide chain against proteolytic attack;
(ii)    the induction and maintenance of the protein conformation in a biologically active form;
(iii)   the decrease of the immunogenicity of proteins;
(iv)    the recognition and association with viruses, enzymes and lectins;
(v)     the control of glycoprotein uptake by cells;
(vi)    intercellular recognition and adhesion.

In addition, we know that the glycan structure of the cell membrane glycoproteins is profoundly altered in cancer cells. This molecular transformation may be related to the appearance of cell surface neoantigens and could be a factor of cancer induction and metastatic diffusion.

Knowledge of the primary structure and conformation of glycans is clearly necessary in order to understand the mechanisms of glycan metabolism and lay the foundations of the molecular biology of glycoproteins.

### 1.1 Types of glycan-protein linkages

Glycans are conjugated to peptide chains through two types of primary covalent linkages ($N$-glycosyl and $O$-glycosyl) leading to the definition of two classes of glycoproteins ($N$-glycosylproteins and $O$-glycosylproteins). The only $N$-glycosidic bond presently found in glycoproteins is $N$-acetylglucosaminyl-asparagine (GlcNAc($\beta$1-$N$)Asn). In contrast, the $O$-glycosidic bond presents a wide variety of linkages, the most common being the following:

(i)     Mucin-type; the alkali-labile linkage between $N$-acetyl-D-galactosamine and L-serine or L-threonine (GalNAc($\alpha$-1,3)Ser or GalNAc($\alpha$-1,3)Thr) found in very numerous glycoproteins which are said to be of the 'mucin type'.

(ii)     Proteoglycan-type; the alkali-labile linkage between D-xylose and L-serine (Xyl($\beta$1-3)Ser) involved in the acidic mucopolysaccharide-protein bond of proteoglycans.

(iii)    Collagen-type; the alkali-stable linkage between D-galactose and 5-hydroxy-D-lysine (Gal($\beta$1-5)OH-Lys) characterized in collagens.

(iv)    Extensin-type; the alkali-stable linkage between L-arabinofuranose and 4-hydroxy-L-proline (l-Ara*f*($\beta$1-4)OH-Pro) identified in plant glycoproteins.

## 1.2 Primary structure of glycoprotein glycans

Glycan structures are not randomly constructed. They may be divided into families with similar structures and common oligosaccharide sequences, whether they originate from animals, plants, microorganisms or viruses. Consequently a number of classes and concepts are firmly established.

(i)      *Concept of the common 'inner-core'*; the carbohydrate moiety of *N*- and *O*-glycosylprotein is derived from the substitution of oligosaccharide structures common to all glycans of a given class of glycoproteins. These non-specific and invariant structures are conjugated to the peptide chain and constitute the most internal part of glycans (the *inner-core, Figure 1*).

(ii)     *Concept of the antenna*; on the basis of their morphology, their flexibility and their property of being recognition signals, the term *antenna* has been proposed for the outer variable arms substituting the inner-cores.

(iii)    *Microheterogeneity of glycans*; in addition to genetic variants expressed as variations in their polypeptide chains, almost all glycoproteins reveal another form of polymorphism associated with their carbohydrate residues. A given glycan located at a given amino acid in a glycoprotein, often presents a structural heterogeneity which is produced by partial substitution of sugar residues on a similar core structure. This type of diversity is termed 'microheterogeneity' or 'peripheral heterogeneity' because it often involves the number and positions of the most externally situated monosaccharides in the glycan. The microheterogeneity is related to variations in the level of sialylation or to more profound modifications like the number of antennae in *N*-glycosylproteins. In all cases, the polymorphism of glycoprotein glycans still poses one of the most formidable problems encountered in the primary structure determination of glycans.

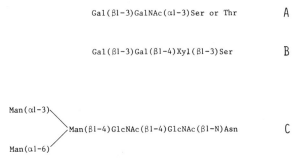

Gal($\beta$1-3)GalNAc($\alpha$1-3)Ser or Thr          A

Gal($\beta$1-3)Gal($\beta$1-4)Xyl($\beta$1-3)Ser          B

Man($\alpha$1-3)⟍
                    ⟍Man($\beta$1-4)GlcNAc($\beta$1-4)GlcNAc($\beta$1-N)Asn          C
Man($\alpha$1-6)⟋

**Figure 1.** Oligosaccharide inner-cores of glycoproteins glycans. Core A exists in the *O*-glycosylproteins of the mucin-type. Core B is the terminal sequence of almost all of the glycosaminoglycans. Core C is common to all *N*-glycosylproteins.

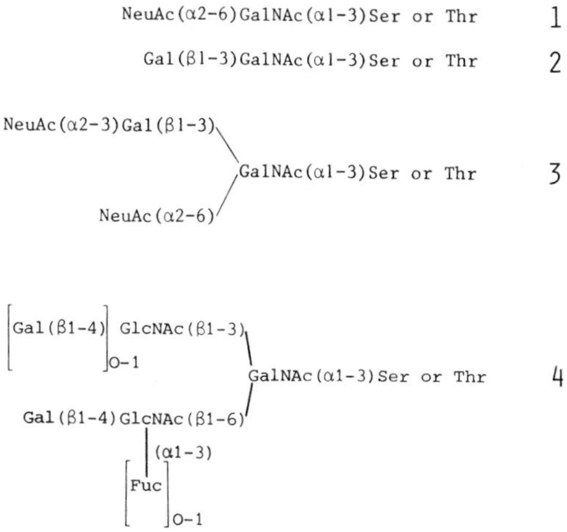

$$\text{NeuAc}(\alpha 2-6)\text{GalNAc}(\alpha 1-3)\text{Ser or Thr} \qquad 1$$

$$\text{Gal}(\beta 1-3)\text{GalNAc}(\alpha 1-3)\text{Ser or Thr} \qquad 2$$

$$\begin{array}{c} \text{NeuAc}(\alpha 2-3)\text{Gal}(\beta 1-3)\diagdown \\ \qquad\qquad\qquad\qquad\quad \text{GalNAc}(\alpha 1-3)\text{Ser or Thr} \qquad 3 \\ \text{NeuAc}(\alpha 2-6)\diagup \end{array}$$

$$\begin{array}{c} \left[\text{Gal}(\beta 1-4)\Big|\ \text{GlcNAc}(\beta 1-3)\diagdown\right._{O-1} \\ \qquad\qquad\qquad\qquad\qquad\quad \text{GalNAc}(\alpha 1-3)\text{Ser or Thr} \qquad 4 \\ \text{Gal}(\beta 1-4)\text{GlcNAc}(\beta 1-6)\diagup \\ \Big|\ (\alpha 1-3) \\ \left[\text{Fuc}\right]_{O-1} \end{array}$$

**Figure 2.** Some examples of structures of glycans *O*-glycosidically conjugated to protein through the linkage GalNAc($\alpha$1-3)Ser or Thr. **(1)** Submaxillary mucins, human erythrocytes. **(2)** Anti-freeze glycoproteins from antarctic fish, human chorionic gonadotropin $\beta$-subunit, human serum IgA, epiglycanin of TA3-Ha cells, T-reactive erythrocytes, rat brain glycoproteins. **(3)** Human glycophorin and gonadotropin $\beta$-subunit, fetuin, bovine kappa-casein, lymphocyte plasma membrane, M-blood group. **(4)** Bronchial mucin of patients suffering from cystic fibrosis. For reviews, see refs. 1–4.

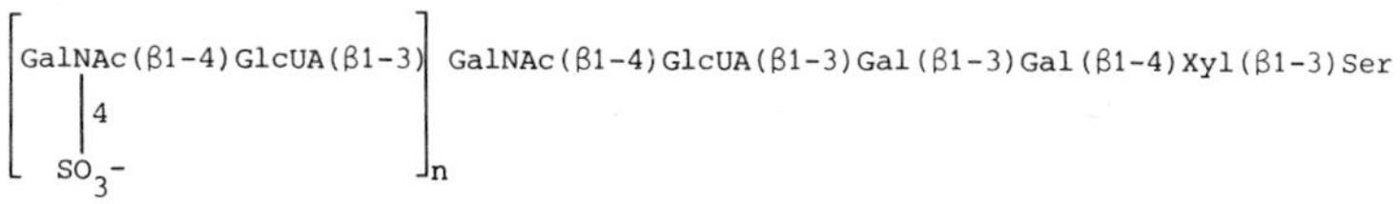

$$\left[\begin{array}{c}\text{GalNAc}(\beta 1-4)\text{GlcUA}(\beta 1-3)\\ \Big|\ 4\\ \text{SO}_3^-\end{array}\right]_n \text{GalNAc}(\beta 1-4)\text{GlcUA}(\beta 1-3)\text{Gal}(\beta 1-3)\text{Gal}(\beta 1-4)\text{Xyl}(\beta 1-3)\text{Ser}$$

**Figure 3.** Primary structure of seryl-chondroitin 4-sulphate.

## 1.2.1 *Primary structure of O-glycosylprotein glycans*

(i) *Glycans conjugated through a GalNAc($\alpha$1-3)Ser or Thr linkage;* this group mainly comprises mucin structures. These glycans are often designated as mucin-like structures, even if they are present in plasma cell membranes and in glycoproteins from biological fluids (*Figure 2*).

(ii) *Glycans conjugated through a Xyl($\beta$1-3)Ser linkage;* this very homogeneous family of glycans consists of the acid mucopolysaccharides (glycosaminoglycans). They are generally linear polymers made up of disaccharide repeating units (*Figure 3*) and *O*-glycosidically linked through the trisaccharide inner-core B of *Figure 1* to the peptide chain in the proteoglycans.

## 1.2.2 *Primary structure of N-glycosylprotein glycans*

The *N*-glycosylproteins are divided into three families according to the nature of the carbohydrate moiety linked to the pentasaccharidic inner-core C (*Figure 1*). In the first family, the glycans contain mannose and *N*-acetylglucosamine only. They are called glycans of the oligomannosidic or high-mannose type (*Figure 4*). In the second family, the sugar composition of glycans is more complex. These glycans contain galactose, fucose and sialic acids in addition to mannose and *N*-acetylglucosamine. They derive

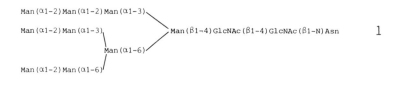

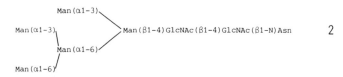

**Figure 4.** Structure of glycans of the oligomannosidic type. Structure 1 (conventionally assigned as $M_9$) is present in calf thyroglobulin unit A, human IgD and myeloma IgM, Chinese hamster ovary cell glycoproteins, bovine lactotransferrin, soybean agglutinin, Newcastle virus and scorpion hemocyanin. Structure 2 (conventionally assigned as $M_5$) has been found in Taka-amylase, hen ovalbumin and human myeloma IgM. For reviews, see refs. 1–4.

fundamentally from the addition to the pentasaccharide inner-core C of a variable number of *N*-acetyllactosamine residues (Gal($\beta$1-4)GlcNAc). These structures have been called glycans of the *N*-acetyllactosaminic or complex type (*Figure 5*). Glycans containing oligo- or poly(*N*-acetyllactosamine) sequences have been found and named poly-(glycosyl)-peptides or poly(*N*-acetyllactosamine)-peptides (*Figure 6*).

In the third family, the glycans contain structures of the oligomannosidic and *N*-acetyllactosaminic types. They belong to the oligomannosido-*N*-acetyllactosaminic or hybrid type (*Figure 7*).

### 1.2.3 *General considerations about the methodology of glycan primary structures*

The problem of the determination of glycan primary structure can be considered as virtually solved due to; (i) the development of efficient methods for the isolation of glycoprotein variants and of their glycan residues, chiefly by using h.p.l.c. and affinity chromatography on immobilized lectins and (ii) the progress in chemical, enzymatic and physical methodologies. In this connection, the collaboration between our laboratory and the one of J.F.G.Vliegenthart led to the introduction of a very efficient and sensitive method for determining the complete primary structure of glycans by associating the permethylation procedure with 360 – 500 MHz $^1$H-n.m.r. spectroscopy. This technique, which requires 25 – 100 $\mu$g of total sugars, is of a general application and is fully described elsewhere (7,8). Recent developments in mass spectrometry, particularly the 'fast atom bombardment' method (f.a.b.), have resulted in considerable progress. The performances of this rapid and sensitive method are detailed in ref. 9. The description of the excellent, but not yet widely applied method of Kobata *et al.*, which associates the permethylation procedure and the stepwise degradation of glycans by glycosidases, has also been omitted due to lack of space (For details, see ref. 11).

In the following paragraphs, we aim to describe the methods we currently use in our laboratory and have tested over numerous years for determining the primary struc-

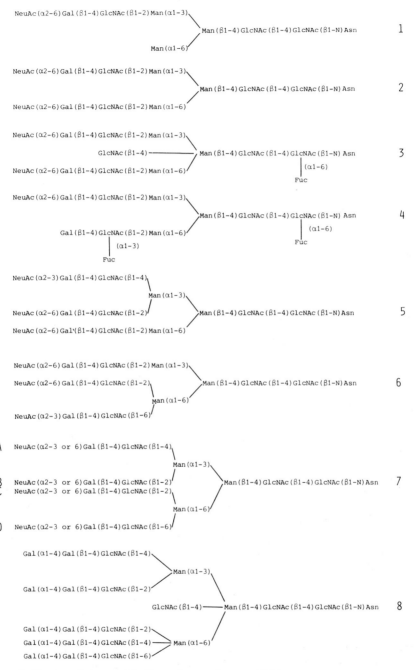

**Figure 5.** Structure of glycans of the *N*-acetyllactosamine type. (**1**) Monoantennary glycan of the secretory component from human milk and human chorionic gonadotropin. (**2**) Biantennary glycan of human serum transferrin. (**3**) Monofucosylated and 'bisected' (presence of a bisecting *N*-acetylglucosamine residue) biantennary glycan of human IgG. (**4**) Difucosylated biantennary glycan of human lactotransferrin. (**5, 6**) Triantennary glycans of human serum transferrin. (**7**) Tetra-antennary glycan of human $\alpha_1$-acid glycoprotein. (**8**) Penta-antennary glycan of turtle-dove ovomucoid. For reviews, see refs. 1−4.

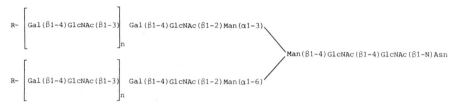

**Figure 6.** General basic structure of the biantennary glycans of the poly (glycosyl)-peptides in which n varies from 1 to 15. In human erythrocyte membrane, R = NeuAc(α2-3)Gal(β1-4)GlcNAc(β1-3), NeuAc(α2-6)-Gal(β1-4)GlcNAc(β1-3), Fuc(α1-2)Gal(β1-4)GlcNAc(β1-3) (Blood group O) or GalNAc(α1-3)[Fuc(α1-2)]-Gal(β1-4)GlcNAc(β1-3) (Blood group A) or Gal(α1-3)[Fuc(α1-2)]Gal(β1-4)GlcNAc(β1-3) (Blood group B).

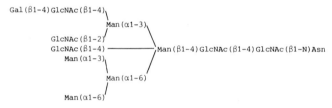

**Figure 7.** Structure of the hybrid type glycan of hen ovalbumin. For reviews, see refs. 1−4.

ture of the glycoprotein glycans. Many of them have been performed by my co-workers, each of whom will describe his own experience in glycan primary structure determination involving (*i*) the isolation of glycoproteins, glycopeptides and glycans, (*ii*) the determination of their composition with respect to simple sugars, (*iii*) the determination of the sequential arrangement of monosaccharide units as well as the linkage between these units and the anomeric configuration of glycosidic bonds, and (*iv*) the determination of the nature of the glycan-peptide linkage.

## 2. CHEMICAL CLEAVAGE OF O- AND N-GLYCOSIDIC LINKAGES OF GLYCANS

The chemical liberation of the carbohydrate moieties of glycoproteins or glycopeptides can be achieved in different ways. The action of alkali in the presence of sodium or potassium borohydride liberates, by a '*β*-elimination' mechanism, the reduced glycans which were *O*-glycosidically linked to serine and threonine. *N*-glycosidically linked glycans can be removed by hydrazinolysis or by alkaline cleavage with total or partial de-*N*-deacetylation. The liberated glycans must be re-*N*-acetylated before further fractionation or investigation.

### 2.1 **Alkaline cleavage of O-glycosidic linkages**

#### 2.1.1 *Principle*

*O*-Glycosidic linkages between glycans and the *β*-hydroxyamino acids serine and threonine are easily split by dilute alkali solution (0.05 − 0.1 M NaOH or KOH) in mild conditions (4 − 45°C for 0.5 − 6 days) by *β*-elimination leading to the liberation of the glycans. In order to prevent the destruction or modification of the latter by a 'peeling reaction', the reaction may be performed in the presence of a reducing agent (0.8 − 2 M NaBH$_4$). Under these conditions, *N*-glycosidic linkages and *O*-glycosidic linkages with hydroxyproline, tyrosine and hydroxylysine are not cleaved.

Concomitantly, under non-reducing conditions, serine and threonine are transformed into dehydro-amino acids (2-aminopropenoic acid and 2-amino-2-butenoic acid, respectively) which strongly absorb u.v.-light at 240 nm. This property is used in order to detect and follow a $\beta$-elimination reaction in non-reductive medium.

The $\beta$-elimination does not occur when the carboxyl group of serine and threonine is free.

### 2.1.2 *Procedure (12)*

*Reductive cleavage of O-glycosidic linkages.*

(i)     Adjust the solution of glycoprotein ($10-20$ mg/ml) to pH 10 with dilute NaOH and add an equal volume of cold freshly prepared 0.1 M NaOH in 2 M NaBH$_4$.
(ii)    Incubate under reflux at 45°C for 16 h.
(iii)   Neutralize the cooled solution by adding 50% acetic acid (until pH 6) and subject the material to gel-filtration on a Sephadex G-50 fine or Bio-Gel P-4 column (75 cm × 2 cm i.d.).

Under these conditions, the *O*-glycosidically linked glycans released as oligosaccharide-alditols are generally well separated from the *N*-glyco-oligopeptides (two to four amino acid residues), if present, formed by alkaline degradation of the peptide chain when both types of linkages co-exist in the studied glycoproteins.

*Fractionation of oligosaccharide-alditols and of N-glycosylpeptides.* Liberated oligosaccharide-alditols can be characterized by thin-layer chromatography (Merck SiO$_2$-plates, solvent-systems: *n*-butanol/acetic acid/water 20:10:15, by volume or *n*-butanol/ethanol/acetic acid/pyridine/water 1:10:0.3:1:3, by volume) and isolated by h.p.l.c. (see Section 3.2). *N*-glycosylpeptides are efficiently fractionated by affinity chromatography on immobilized lectins (see Section 4.4).

## 2.2 **Akaline cleavage of *N*-glycosidic linkages (13)**

### 2.2.1 *Principle*

Although *N*-glycosidic linkages are stable in the mild alkaline conditions of $\beta$-elimination, they are quantitatively cleaved under drastic alkaline conditions (1 M NaOH or saturated barium hydroxide at 100°C for $6-24$ h). Under these conditions, (i) alkaline attack must be performed in a reducing medium ($1-2$ M NaBH$_4$) in order to prevent the 'peeling reaction' of liberated *N*-glycosidically linked glycans but without avoiding the profound degradation of *O*-glycosidically linked glycans; and (ii) *N*-acetylglucosamine residues are deacetylated so that, in a second step, liberated glycans have to be re-*N*-acetylated with acetic anhydride (sometimes, radiolabelled).

### 2.2.2 *Procedure*

(i)     Dissolve the *N*-glycosylprotein (50 mg/ml) in a 1 M NaOH − 1 M NaBH$_4$ solution.
(ii)    Heat under reflux for 6 h at 100°C.

(iii)    Neutralize the mixture, maintained in an ice-bath, with 50% acetic acid to a pH value of 6.

(iv)    Purify the material on a Bio-Gel P-2 column (40 cm × 2 cm i.d.), pool the carbohydrate containing fractions and lyophilize.

(v)    In order to re-*N*-acetylate, dissolve the material in saturated $NaHCO_3$ (1 ml/mg of glycan), add five aliquots of 10 $\mu$l of acetic anhydride at 5 min intervals, desalt the reaction mixture on a Dowex 50 × 8 (25 − 50 mesh; $H^+$) column (5 cm × 1 cm i.d.) and lyophylize the effluent.

(vi)    Fractionate the glycan-alditols either by h.p.l.c. (see Section 3.2) or by affinity chromatography on immobilized lectins (see Section 4).

## 2.3 Hydrazinolysis of *N*-glycosidic linkages (14)

### 2.3.1 *Principle*

Hydrazine cleaves the *N*-glycosidic linkages of *N*-glycosylpeptides and *N*-glycosylproteins. It liberates the *N*-deacetylated glycans as their hydrazones. Thus the procedure for obtaining the carbohydrate moieties is carried out in three steps: hydrazinolysis, re-*N*-acetylaion and reduction in order to stabilize the molecule.

### 2.3.2 *Procedure*

(i)    In a screw cap tube with a teflon-silicone disc seal, introduce the dried glycopeptide or glycoprotein plus the minimum amount of anhydrous hydrazine (Pierce Chemical Co., Rockford, IL, USA) necessary in order to cover the sample.

(ii)    Heat at 100°C for 12 − 14 h. Do not stir using a Vortex, which deposits the material on the wall of the tube.

(iii)    Eliminate the excess of hydrazine by repeated evaporation in presence of toluene (1 vol) under a stream of nitrogen. Remove the last traces of hydrazine in a vacuum dessicator over $H_2SO_4$.

(iv)    Purify the material on a Bio-Gel P-2 column (40 cm × 2 cm i.d.), pool the carbohydrate fractions and lyophilize.

(v)    Re-*N*-acetylation is carried out as described in Section 2.2.2, step (v).

(vi)    Dissolve the oligosaccharide fraction in 0.05 M NaOH (1 ml/mg), reduce with $NaBH_4$ (5 mg/mg of sugar) for 16 h at 20°C and stop the reaction by adding Dowex 50 × 8 (25 − 50 mesh; $H^+$).

(vii)    Filter, wash the filter with water and concentrate the filtrate *in vacuo*.

(viii)    Eliminate boric acid by repeated co-distillation *in vacuo* with methanol.

(ix)    Purify the material by gel-filtration on a Bio-Gel P-2 column (25 cm × 2 cm i.d.).

(x)    Glycans − alditols are fractionated as described in Section 3.2.2, step (ii).

### 2.3.3 *General comments*

(i)    Radiolabels may be introduced by re-*N*-acetylation with radiolabelled acetic anhydride, or reduction with sodium borotritide. This procedure is currently applied in our laboratory for fractionating the cell membrane glycoprotein glycans by affinity chromatography on immobilized lectins (see Section 4.3).

(ii)    The *N*-acetylglucosamine residue involved in the glycosylamine linkage is converted into the hydrazone derivative. This leads to the formation of by-products in the course of the following reactions:

(a) The re-*N*-reacetylation step gives rise to 1-acetylhydrazo-*N*-acetyl-glucosamine, 60% of which is converted, during the treatment with the cationic resin, into *N*-acetylglucosamine which will be transformed during the reduction into *N*-acetylglucosaminitol, while the remaining 40% of the hydrazone derivative is reduced into 1-*N*-acetylhydrazino-*N*-acetyl-glucosaminitol.

(b) The latter will be converted, in case of permethylation, into a mixture of permethylated epimeric 1-deoxyalditols.

(c) Other minor products, such as 1-deoxy-*N*-acetylglucosaminitol and 1-di-*N*-acetylhydrazino-*N*-acetylglucosaminitol are also present as final products of reduction and are responsible of the complex pattern of *N*-acetyl signals observed by n.m.r. spectroscopy.

(iii) Incomplete de-*N*-acetylation of C-3 substituted *N*-acetylglucosamine has been reported and must be taken into consideration when quantitative re-*N*-acetylation with radiolabelled acetic anhydride is performed.

(iv) The behaviour toward hydrazine of *O*-glycosidically linked glycans has not yet been elucidated. However, some results show that they are profoundly degraded until an alkali stable linkage stops the 'peeling' reaction.

## 2.4 Production of glycans from O,N-glycosylproteins

When glycosylproteins possess both *O*- and *N*-glycosidically linked glycans, we use two procedures: (i) successive use of reductive β-elimination and alkaline cleavage, and (ii) combining reductive β-elimination with hydrazinolysis.

### 2.4.1 *Successive β-elimination and alkaline cleavage*

Only the *O*-glycosidically linked glycans are liberated, in the first step, by reductive-mild alkaline treatment as oligosaccharide-alditols, which are stable in the drastic conditions applied in a second step in order to split the *N*-glycosidic linkages.

(i) Incubate the glycoprotein or glycopeptide under reflux at 45°C for 16 h in 0.05 M NaOH − 1 M NaBH$_4$ as described in Section 2.1.2.

(ii) Add an equal volume of 2 M NaOH − 1 M NaBH$_4$ and heat again under reflux for 6 h at 100°C.

(iii) Neutralize the solution and purify the carbohydrate material as described in Section 2.1.2 and re-*N*-acetylate and fractionate according to Section 2.2.2.

### 2.4.2 *Successive β-elimination and hydrazinolysis*

*O*-glycosidically linked glycans are first liberated as oligosaccharide-alditols by β-elimination and separated by gel-filtration from the *N*-glycosylpeptides which are hydrazinolysed separately later.

(i) Incubate the glycoprotein or glycopeptide under reflux at 45°C for 16 h in 0.05 M NaOH − 1 M NaBH$_4$ as described in Section 2.1.2.

(ii)     Neutralize the solution and purify the carbohydrate material as described in Section 2.1.2.

(iii)    Collect the *N*-glycosylpeptides and oligosaccharide-alditols and lyophilize.

(iv)     Submit the *N*-glycosylpeptides to the hydrazinolysis procedure (see Section 2.3.2).

(v)      Fractionate the glycan-alditols as described in Section 2.2.2. If the oligosaccharide-alditols and *N*-glycosylpeptides are not well fractionated by gel filtration, submit both to the hydrazinolysis.

## 3. GLYCOPEPTIDE AND GLYCAN ISOLATION

The preparation of glycoproteins is not described in this chapter because these compounds cannot be isolated by a uniform procedure, due to the variability in amount and properties of the carbohydrate moieties they carry. In general, the techniques applied are the classical methods for isolating proteins (for review, see ref. 20). However, an additional and efficient procedure has been developed in the last few years which is based on the use of lectins and which takes advantage of the presence of the carbohydrate part (see Section 4.3).

In contrast, fractionation and isolation of glycans and glycopeptides is relatively easy, although the mixtures are often complex due to microheterogeneity.

### 3.1 Isolation of glycopeptides

#### 3.1.1 *Proteolysis of glycoproteins*

Glycopeptides and glyco-amino acids are obtained by the action of various proteases, the most efficient one being Pronase from *Streptomyces griseus*.

(i)      To 100 mg of glycoprotein dissolved in 10 ml of 0.01 M calcium acetate, add 2 mg of Pronase E (Merck, Darmstadt, FRG).

(ii)     Incubate the mixture at 40°C for 24 h. After 4, 8 and 20 h, add the same amount of Pronase.

(iii)    Maintain a constant pH 8 by continuous addition of a solution of 0.1 M NaOH (use a pH-stat). After 24 h, adjust the proteolytic digest to pH 4.5 with dilute acetic acid and concentrate to 1 ml.

(iv)     Add 10 ml of cold ($-10°C$) ethanol. Leave the mixture at room temperature for 2 h and then at 2°C overnight.

(v)      Centrifuge and discard the precipitate.

#### 3.1.2 *Purification of glycopeptides*

*Purification by ion-exchange chromatography.*

(i)      Add an equal volume of a 10% (w/v) aqueous solution of trichloroacetic acid to the solution obtained in Section 3.1.1 step (v).

(ii)     Centrifuge and desalt the supernatant by passing through coupled columns of Dowex 50 × 8 (25−50 mesh, H$^+$) and Duolite A 102D (25−50 mesh, OH$^-$).

(iii)    Wash the columns with water and concentrate the eluates under vacuum.

*Purification by gel-filtration:* Remove the large peptides by a gel-filtration on Sephadex G-25 or Bio-Gel P-2 eluting with water or 5% (v/v) acetic acid.

### 3.1.3 *Fractionation of glycopeptides*

*Gel-filtration chromatography.*

(i)    Equilibrate two coupled columns (140 cm × 1.8 cm i.d.) of Bio-Gel P-4 in 1% (v/v) acetic acid at room temperature.

(ii)    Elute the glycopeptides at a flow rate of 5 ml/h and collect fractions of 1 ml.

(iii)    Analyse each fraction by paper electrophoresis (see below) or by t.l.c. on Silica-gel G-60 using *n*-butanol/ethanol/water/acetic acid/pyridine (10:100:30:3:10 by volume) as solvent.

(iv)    After 5 h, stain the Silica-gel plates with a reagent containing 0.2% (w/v) orcinol in 20% $H_2SO_4$ (v/v) and heat at 110°C.

*Ion-exchange chromatography.*

(i)    Anion-exchange chromatography: perform the ion-exchange chromatography of glycopeptides (40 mg) on a Dowex 1 × 2 (200 – 400 mesh) column (36 cm × 2.5 cm i.d.), equilibrated with 2 mM pyridinium acetate, pH 5. Elute the neutral glycopeptides with the starting buffer, monosialylated glycopeptides with 50 mM pyridinium acetate (pH 5), disialylated glycopeptides with 500 mM pyridinium acetate (pH 5). Monitor the glycopeptides with the phenol-sulfuric reagent described in Chapter 1, Section 2.1.2. Concentrate each fraction under vacuum, lyophilize and purify by gel filtration on a Bio-Gel P-2 column.

(ii)    Cation-exchange chromatography: fractionate the glycopeptides (40 mg) on a DEAE-Sephadex A 50 column (42 cm × 1.6 cm i.d.) equilibrated with 2 mM pyridinium acetate, pH 5. Elute the glycopeptides, first with the starting buffer and then with a linear gradient from 2 mM to 500 mM of pyridinium acetate, pH 5. Monitor and purify the fractions as described.

*High-performance liquid chromatography:* h.p.l.c. of glycopeptides is not yet widely developed because of the interference from the peptide moiety. In fact, a given glycan located in a given peptide sequence of the protein generally gives rise to a mixture of glycopeptides due to the random nature of proteolytic action. This heterogeneity is in addition to the microheterogeneity of glycans. In this connection, chemical or enzymatic removal of glycans from such complex mixtures solves this problem and allows the use of h.p.l.c. (see Section 3.2).

*Affinity chromatography on immobilized lectin:* affinity chromatography on immobilized lectins (see Section 4.4) constitutes one of the best tools for isolating glycopeptides since the peptide moiety does not interfere in the binding to the lectin.

*Preparative electrophoresis:*

(i)    Free-flow electrophoresis. Use an Elphor-Vap apparatus and fractionate the glycopeptides in 1 M acetic acid, pH 2.4, at 1700 V.

(ii)    High-voltage electrophoresis. Perform high voltage electrophoresis on Whatman 3 paper at pH 6.4 (pyridinium acetate), for 2 h, at 4000 V.

(iii)    Low voltage electrophoresis. Carry out overnight the low voltage electrophoresis on Whatman 3 paper at pH 2.4 (1 M acetic acid) or at pH 3.9 (pyridine/acetic acid/water, 15:50:1935 by volume).

*Staining of glycopeptides:* Detection of glycopeptides in the fractions can be carried out by depositing a drop of each of the fractions (on sheets of Whatman paper) and by applying to the dried spots a colour generating reaction specific to either the peptides or sugar moiety.

*Peptide detection.* Dip the Whatman paper sheet in a ninhydrin reagent prepared as follows: to a solution of 1 g of cadmium acetate in a mixture of glacial acetic acid (50 ml) and water (100 ml), add a solution of 10 g of ninhydrin in acetone and make up to 1 l with more acetone. Develop the violet colour by heating the paper at 150°C.

*Carbohydrate detection.* Dip the sheet of Whatman paper in a solution of 0.1 M periodic acid in acetone. Dry at room temperature and dip the sheet in a 0.15% (w/v) of benzidine in acetone. Sugars develop a yellow colour on a green background.

## 3.2 Isolation of glycans by h.p.l.c.

Glycans can be fractionated and isolated by applying the classical methods of electrophoresis and of adsorption-, partition- and ion-exchange chromatography. However, these procedures have recently been gradually replaced by the more sensitive and rapid methods of h.p.l.c. and of affinity chromatography on immobilized lectins (see Section 4). Therefore, we shall restrict ourselves to the description of the procedures of glycan isolation by h.p.l.c.

H.p.l.c. is widely applied to the separation and preparation of oligosaccharides liberated from glycopeptides by alkaline cleavage (see Sections 2.1 and 2.2), hydrazinolysis (see Section 2.3) or hydrolysis by endoglycosidases (see Section 8.3). We currently use in our laboratory the following h.p.l.c. techniques:

(i)     anion-exchange chromatography of sialyloligosaccharides;
(ii)    partition chromatography of neutral and acidic oligosaccharides on primary amine-bonded silica or alkyl diol-bonded silica;
(iii)   reverse-phase chromatography of neutral oligosaccharides on $C_2$ and $C_{18}$-bonded silica.

### 3.2.1 *Preparative anion exchange h.p.l.c. of sialyloligosaccharides on Micropak AX-10 anion exchange column (21)*

Perform h.p.l.c. of sialyloligosaccharides liberated by reductive alkaline cleavage of *O*- and *N*-glycosylpeptides (see Sections 2.1 and 2.2) or by hydrazinolysis of *N*-glycosylpeptides (see Section 2.3), on a 10 $\mu$ Micropak AX-10 column (50 cm × 0.8 cm i.d.; Varian Associates, Walnut Creek, CA, USA) with a Spectra-Physics liquid chromatograph (Model 8.700) equipped with Model 8400 variable-wavelength detector connected to a Model 4100 computing integrator (Spectra-Physics Inc., San Jose, CA, USA).

(i)     Dissolve 20 mg of the oligosaccharides in 60 $\mu$l of distilled water and submit to h.p.l.c. using a gradient of 0.5 M $KH_2PO_4$ adjusted to pH 4.0 with phosphoric acid under the following conditions: stepwise elution with distilled water for 5 min, linear gradient to 2.5% (v/v) 0.5 M $KH_2PO_4$ for 15 min, isocratic elution for 25 min, linear gradient to 5% (v/v) 0.5 M $KH_2PO_4$ for 35 min, isocratic elution for 45 min, and finally linear gradient to 40% (v/v) 0.5 M $KH_2PO_4$ for 40 min. The flow rate throughout should be maintained at 2 ml/min. Oligosaccharides are detected at 200 nm with detector sensitivity 0.32 and integrator attenuation 16 using an integrator chart speed of 0.5 cm/min.

(ii)  Purify each fraction (4 ml) by gel filtration on Bio-Gel P-2 (200 – 400 mesh) column (30 cm × 1.9 cm i.d.), the elution being carried out with water at a flow rate of 12 ml/h.

(iii)  Detect the sugars present in each fraction with orcinol-sulphuric acid reagent (see Chapter 1, *Table 3*) on Silica Gel plates (precoated Silica Gel-60; Merck, Darmstadt, FRG).

(iv)  Detect the phosphate salts eluted from the Bio-Gel column by precipitation with silver nitrate (see *Figure 8* for an example).

### 3.2.2 *Partition h.p.l.c. of neutral and acidic oligosaccharides on primary amine-bonded silica*

(i) *Partition h.p.l.c. on bonded primary amine packings of neutral oligosaccharides obtained by hydrazinolysis of N-glycosylpeptides:* Carry out analysis and preparative chromatography of glycans liberated by hydrazinolysis (see Section 2.3) of non-sialylated *N*-glycosylpeptides with a Spectra-Physics apparatus Model 8700, equipped with an u.v. 8400 variable wavelength detector connected to a 4100 computing integrator (Spectra-Physics Inc., San Jose, CA, USA). Perform the chromatography on a 5 $\mu$ Amino AS-5A (Brownlee Labs, Santa Clara, CA, USA) column (25 cm × 4 mm i.d.)

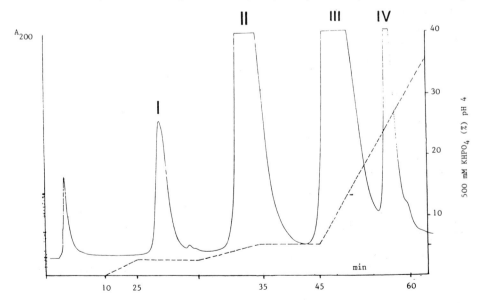

**Figure 8.** H.p.l.c. on 10 $\mu$ Micropak AX-10 column of sialoglycans liberated by hydrazinolysis of $\alpha_1$-acid glycoprotein. I, II, III, IV: mono-, di-, tri- and tetrasialylated glycans. The recovery was 91%.

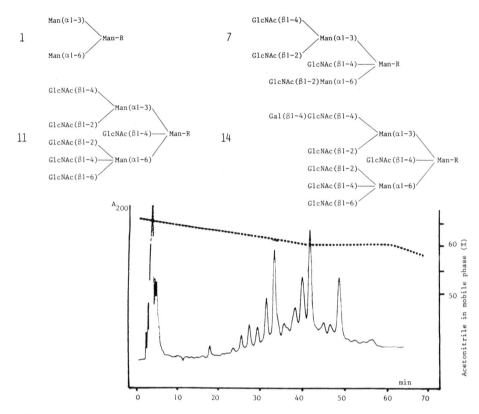

**Figure 9.** H.p.l.c. on 5 $\mu$ Amino AS-5A of glycans liberated by hydrazinolysis of hen ovomucoid. Primary structure of glycans 1, 7, 11 and 14 are given above the elution diagram. R = ($\beta$1-4)GlcNAc($\beta$1-4)GlcNAc-ol.

equilibrated with acetonitrile-water (65/35, v/v). Inject a solution of 5 mg of oligosaccharides dissolved in 30 $\mu$l of acetonitrile-water (50/50, v/v) and apply the following stepwise elution conditions: linear gradient from acetonitrile-water (65/35, v/v) to acetonitrile-water (60/40, v/v) for 30 min followed by isocratic conditions for 30 min and then a linear gradient to acetonitrile-water (50/50 v/v) for 30 min. The flow-rate should be 1 ml/min. Oligosaccharides are detected at 200 nm with detector sensitivity 0.32 and integrator attenuation 50.

The separation obtained within 90 min of a mixture of 17 reduced oligosaccharides liberated from hen ovomucoid by hydrazinolysis is shown in *Figure 9*.

(ii) *Partition h.p.l.c. on bonded primary amine packings of sialyloligosaccharides obtained by alkaline borohydride treatment of O-glycosylpeptides (22):* Analyse the oligosaccharide-alditols obtained by $\beta$-elimination from *O*-glycosylpeptides (see Section 2.1) with the same Spectra-Physics material as described above. Dissolve 1 mg of oligosaccharides in 10 $\mu$l of distilled water and inject into a 5 $\mu$ Amino AS-5A (Brownlee Labs, Santa Clara, CA, USA) column (25 cm × 4 mm i.d.) equilibrated with the initial solvent (acetonitrile-phosphate buffer 15 mM, pH 5.2; 4/1 v/v). Develop the chromatogram isocratically for 25 min with 4/1 (v/v) acetonitrile-15 mM phosphate

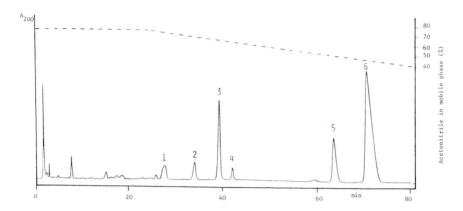

**Figure 10.** H.p.l.c. on 5 $\mu$ Amino-5A of glycan-alditols liberated by $\beta$-elimination from Cad erythrocyte membrane glycophorin. 1: NeuAc($\alpha$2-3)Gal($\beta$1-3)GalNAc-ol; 3: NeuAc($\alpha$2-3)GalNAc($\beta$1-4)Gal($\beta$1-3)-GalNAc-01; 5: NeuAc($\alpha$2-3)Gal($\beta$1-3)[NeuAc($\alpha$2-6)]GalNAc-ol; 6: NeuAc($\alpha$2-3)[GalNAc($\beta$1-4)]Gal($\beta$1-3)-[NeuAc($\alpha$2-6)]GalNAc-ol.

buffer (pH 5.2), after which apply a linear gradient to decrease the acetonitrile concentration in the mobile phase. Use a flow rate of 1 ml/min. Oligosaccharides are detected at 200 nm with detector sensitivity 0.35 and integrator attenuation 4.

An example of separation of six oligosaccharides liberated from Cad glycophorin A by $\beta$-elimination is given in *Figure 10*.

### 3.2.3 *Partition h.p.l.c. of oligosaccharides of the oligomannosidic type on alkyl diol-bonded silica*

Oligosaccharides of the oligomannosidic type (see *Figure 4*) liberated from *N*-glycosylpeptides by hydrazinolysis (see Section 2.3), by alkaline cleavage (see Section 2.1 and 2.2) or by endoglycosidases (see Section 8.3.3), as well as oligomannosides present in the urine of patients with mannosidosis are easily separated by partition h.p.l.c. on alkyl diol-bonded silica.

The procedure performed in our laboratory is as follows: we use the Spectra Physics apparatus as described above.

(i)     Equilibrate the column with the initial solvent (acetonitrile-water, 7/3 v/v).

(ii)    Inject into a 3 $\mu$ Lichrosorb diol column (Merck, Darmstadt, FRG) 1 mg of oligosaccharides dissolved in 20 $\mu$l of distilled water.

(iii)   Chromatograph with a linear gradient to acetonitrile-water (60/40 v/v) for 50 min at a flow rate 1 ml/min. Oligosaccharides are detected at 200 nm with detector sensitivity 0.5 and integrator attenuation 4.

*Figure 11* shows an example of separation of eight oligomannosides.

### 3.2.4 *Reverse-phase h.p.l.c. of neutral oligosaccharides on $C_{18}$-bonded silica (23)*

Perform h.p.l.c. on a 3 $\mu$ $C_{18}$-column (10 cm $\times$ 4.6 mm i.d.; Brownlee Labs, Santa Clara, CA, USA) using a Spectra-Physics apparatus as described above. Inject into the column 1 mg of oligosaccharides dissolved in 10 $\mu$l of distilled water. Chromatograph

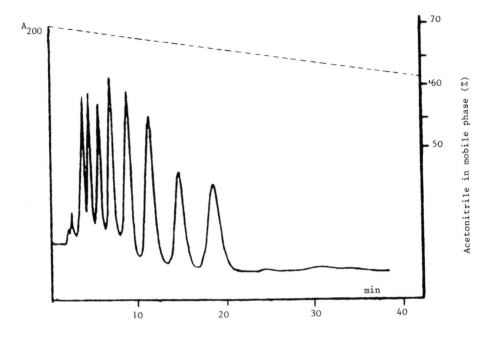

**Figure 11.** H.p.l.c. on alkyl diol-bonded silica of oligomannoside-alditols from urine of mannosidosis. $M_2G$-ol to $M_9G$-ol: oligomannoside-alditols containing from 2 to 9 mannose residues.

by isocratic elution with distilled water for 15 min, followed by a linear gradient to water-acetonitrile (90/10 v/v) for 25 min after which apply isocratic conditions with the same solvent for 20 min. Use a flow rate of 1 ml/min. Oligosaccharides are detected at 200 nm with detector sensitivity 0.8 and integrator attenuation 8.

As an example, from fraction 9 shown in *Figure 9*, 15 fractions were obtained by reverse-phase chromatography. This indicates the high degree of microheterogeneity of glycans of hen ovomucoid and the resolution power of the method (*Figure 12*).

## 4. USE OF IMMOBILIZED LECTINS

### 4.1 **Introduction to the use of lectins**

Lectins are 'sugar-binding proteins or glycoproteins of non-immune origin which agglutinate cells and/or precipitate glycoconjugates'. They bear at least two sugar-binding sites, the presence of which explains the ability of lectins to precipitate polysaccharides, glycoproteins and glycolipids and to agglutinate cells. As these interactions are often reversed by monosaccharides or glycosides, lectins are powerful tools for isolating sugar components and cells either in their soluble or immobilized form (for reviews, see ref. 24 and 25).

The carbohydrate specificity of a given lectin is usually established by the method of fixation-site saturation in which different sugars are tested for their ability to inhibit

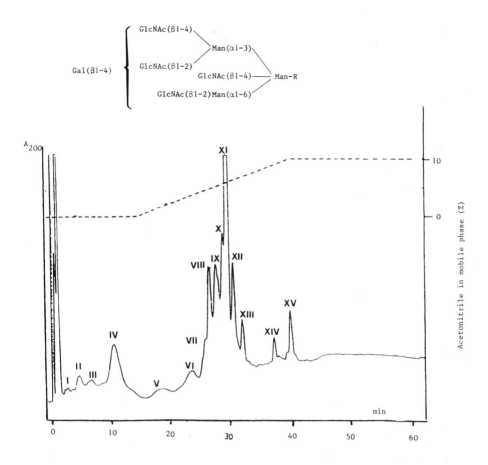

**Figure 12.** Reverse-phase h.p.l.c. on a 3 μ C$_{18}$-column of peak 9 from Figure 9 obtained after preparative h.p.l.c. of the glycan-alditols liberated from hen ovomucoid by hydrazinolysis. Only the structure of oligosaccharide XI is given in the Figure. R: (β1-4)GlcNAc(β1-4)GlcNAc-ol.

either haemagglutination or precipitation of polysaccharides and/or glycoconjugates by the lectin. Another procedure consists of the determination of the association constant by equilibrium analysis.

The specificity of a lectin is defined in terms of the best monosaccharide inhibitor. The results from such studies lead to the classification of lectins into groups some of which are given in *Table 1*.

Originally, it was thought that lectins recognized monosaccharides in the non-reducing external position. However, it was further demonstrated that in most cases complex oligosaccharides are several thousand-fold more potent as 'haptens' toward lectins than monosaccharides themselves so that the concept of the 'lectino-dominant monosaccharide' must be replaced by that of the 'lectino-dominant oligosaccharide' (28 – 31). Moreover, we know that the lectins are able to bind 'laterally' to glycoprotein glycans.

159

**Table 1.** Lectins commonly used for glycoprotein study and classified according to their monosaccharide specificity

| | |
|---|---|
| $\alpha$-D-Mannose, $\alpha$-D-Glucose | |
| *Canavalia ensiformis* | Con A |
| *Lens culinaris* | LCA |
| $\beta$-D-Galactose, *N*-acetyl-$\beta$-D-galactosamine | |
| *Ricinus communis* | $RCA_I$ |
| | $RCA_{II}$ |
| *Glycin max* (Soybean) | SBA |
| *Arachis hypogaea* (Peanut) | PNA |
| $\alpha$-D-Galactose, *N*-acetyl-$\alpha$-D-galactosamine | |
| *Griffonia simplicifolia I* | $GSA_I$ |
| *Dolichos biflorus* | DBA |
| *N*-Acetyl-$\beta$-D-glucosamine | |
| *Triticum vulgare* (Wheat germ) | WGA |
| $\alpha$-L-Fucose | |
| *Lotus tetragonolobus* | LTA |
| *Ulex europeus I* | $UEA_I$ |
| $\alpha$-*N*-Acetylneuraminic acid | |
| *Limulus polyphemus* (Limulin) | |

The structures which are actually recognized by lectins commonly used in biochemical and biological research are described in *Table 2*. In addition, we must keep in mind that the affinity of immobilized lectins toward free and/or conjugated sugars is absolutely unforseeable and cannot be deduced at all from the results obtained with the lectins in solution from haemagglutination experiments (28,29). For example, oligosaccharides are able to inhibit the haemagglutination by LCA, (see *Tables 1* and *2* for lectin abbreviations) but only glycoasparagines, glycopeptides and glycoproteins are fixed on the immobilized LCA. So, oligosaccharides liberated by hydrazinolysis or alkaline cleavage are not retained by this latter material. In the same way, the presence of the 'bisecting' *N*-acetylglucosamine residue (see *Figure 5.3*) does not influence the inhibitory effect of oligosaccharides on the haemagglutination by Con A, but hinders the fixation of glycans on immobilized Con A (31).

In conclusion, the only method for determination of the binding specificity of any immobilized lectin consists of the study of the behavior of free and conjugated oligosaccharides of well defined primary structure such as those listed in *Table 3*.

## 4.2 **Immobilization of lectins**

Numerous methods have been proposed for preparing immobilized lectins, many of which are now available from different suppliers (Pharmacia, Miles, E.Y. Laboratories, I.B.F.). Lectins directly coupled to CNBr-activated agarose are the most popular. However, immobilized lectins can be easily prepared using agarose, Sepharose 4B or Ultrogel, at a concentration of $2-10$ mg of lectin per ml of settled gel. Immobilization and subsequent use of the immobilized lectins requires some essential rules.

(i) During the coupling reactions, lectin binding sites must be protected by addition of the monosaccharide specific of the lectin (hapten sugar).

(ii) In some cases, non-specific (hydrophobic or ionic) adsorption of glycoproteins

**Table 2.** Specificity of commonly used lectins towards oligosaccharide sequences belonging to *N*-glycosylproteins.

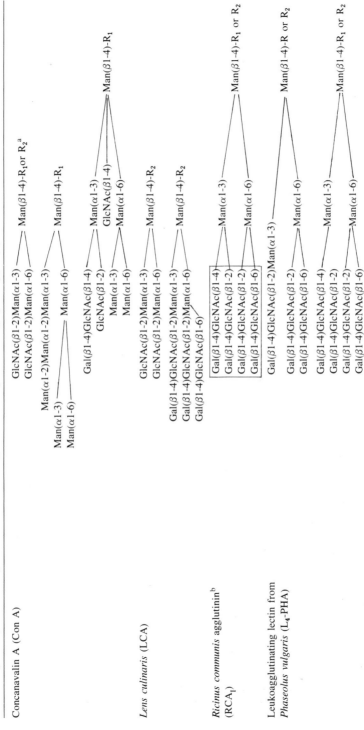

**Table 2.** continued.

Wheat germ agglutinin (WGA)[b]

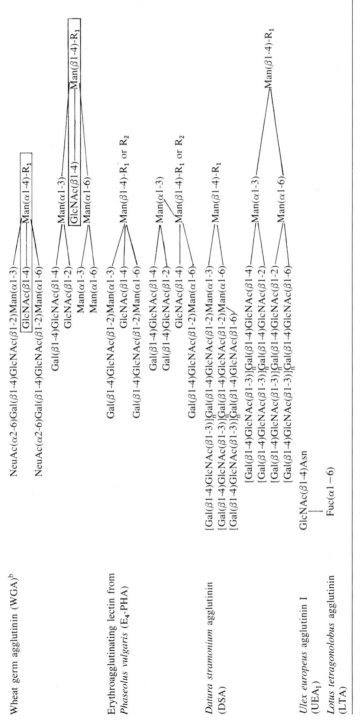

NeuAc(α2-6)Gal(β1-4)GlcNAc(β1-2)Man(α1-3)

GlcNAc(β1-4)

Man(α1-4)-R₁

NeuAc(α2-6)Gal(β1-4)GlcNAc(β1-2)Man(α1-6)

GlcNAc(β1-4)

Man(β1-4)-R₁

Gal(β1-4)GlcNAc(β1-4)

GlcNAc(β1-2)

Man(α1-3)

Man(α1-6)

Erythroagglutinating lectin from *Phaseolus vulgaris* (E₄-PHA)

Gal(β1-4)GlcNAc(β1-2)Man(α1-3)

GlcNAc(β1-4)

Man(β1-4)-R₁ or R₂

Gal(β1-4)GlcNAc(β1-2)Man(α1-6)

Gal(β1-4)GlcNAc(β1-4)

Gal(β1-4)GlcNAc(β1-2)

Man(α1-3)

GlcNAc(β1-2)Man(α1-6)

Man(β1-4)-R₁ or R₂

Gal(β1-4)GlcNAc(β1-2)Man(α1-6)

*Datura stramonium* agglutinin (DSA)

[Gal(β1-4)GlcNAc(β1-3)]ₙGal(β1-4)GlcNAc(β1-2)Man(α1-3)

[Gal(β1-4)GlcNAc(β1-3)]ₙGal(β1-4)GlcNAc(β1-2)Man(α1-6)

Man(β1-4)-R₁

[Gal(β1-4)GlcNAc(β1-3)]ₙGal(β1-4)GlcNAc(β1-6)

[Gal(β1-4)GlcNAc(β1-3)]ₙGal(β1-4)GlcNAc(β1-4)

[Gal(β1-4)GlcNAc(β1-3)]ₙGal(β1-4)GlcNAc(β1-2)

[Gal(β1-4)GlcNAc(β1-3)]ₙGal(β1-4)GlcNAc(β1-2)

Man(α1-3)

Man(α1-6)

Man(β1-4)-R₁

*Ulex europeus* agglutinin I (UEA₁)

GlcNAc(β1-4)Asn

|

Fuc(α1-6)

*Lotus tetragonolobus* agglutinin (LTA)

[a] R₁: GlcNAc(β1-4)GlcNAc(β1-N) Asn; R₂: GlcNAc(β1-4)[(Fuc(α1-6)]GlcNAc(β1-N)Asn;
[b] Sequences in boxes are the minimal oligosaccharide structure necessary for lectin recognition.

**Table 3.** Structure of *N*-glycosylpeptides used for studying immobilized lectin specificity.

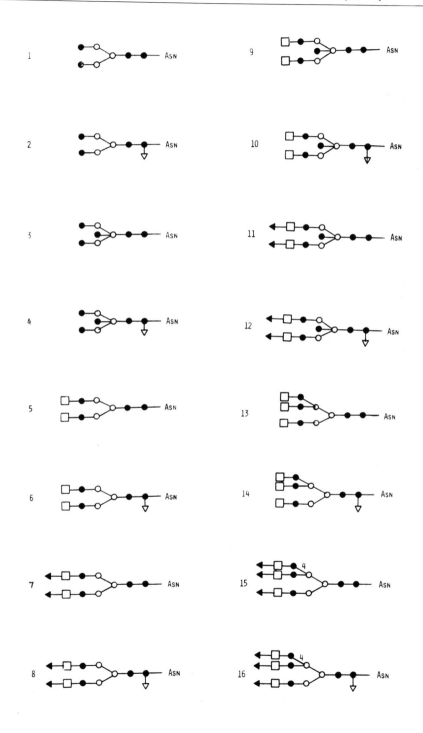

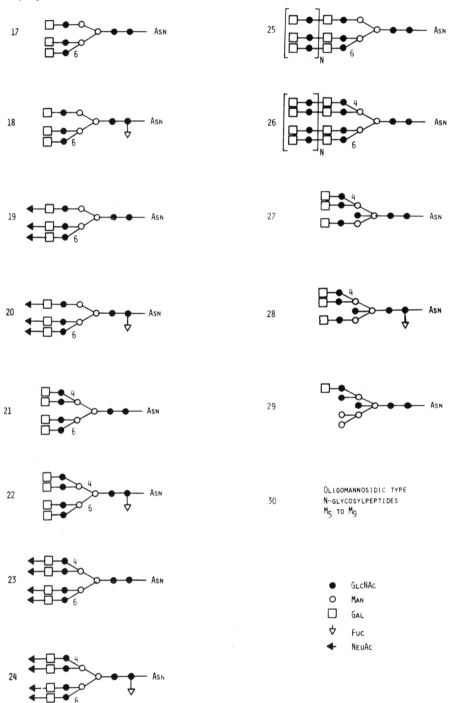

Some of the structures are detailed in *Figures 4–7* and in *Table 2*. Numbers given in the schemes refer to glycosidic linkages: 4: β-1,4; 6: β1-6. In all schemes, the antennae linked to the α-1,3 and α-1,6 mannose residues are the upper and the lower ones, respectively.

on the immobilized lectins has been observed. In order to avoid this phenomenon, new types of immobilization matrices of lectins have been developed, such as polyacrylic-hydrazido-agarose on which lectins can be coupled with glutaraldehyde, giving a stable, non-leaking, uncharged and hydrophilic absorbent.

(iii) Immobilized lectins must be stored at 4°C in buffer containing 0.02% sodium azide as a bacteriostatic agent, in order to avoid any loss of activity for several years.

(iv) Some lectins possess metal-binding sites and the presence of metal ions is important to induce a proper conformation of the lectin needed for carbohydrate binding. For example, Con A needs $Mn^{2+}$ and $Ca^{2+}$ for full activity so that the buffers used for the affinity chromatography on this lectin must contain $MnCl_2$ and $CaCl_2$ (1 mM each).

(v) In order to limit non-specific interactions betweeen glycoproteins and immobilized lectins, such as ionic interactions, the buffers must possess a moderate ionic strength $(0.1 - 1$ M in NaCl).

### 4.2.1 *Activation of Sepharose 4B with CNBr according to March et al. (32) and coupling of lectins*

(i) Wash 50 ml of wet Sepharose 4B beads (Pharmacia) on a sintered glass filter with 1 l of chilled distilled water and transfer into a beaker cooled in an ice-bath.

(ii) Mix 100 ml of ice cold 2 M $K_2CO_3$ solution with the wet filtered Sepharose beads, under gentle magnetic stirring. Add 6.25 ml of a freshly prepared CNBr solution in acetonitrile (2 g/ml) and maintain at 0°C for 2 min in a well ventilated hood.

(iii) Quickly pour the entire reaction mixture into a sintered glass filter funnel and wash extensively with 3 l of ice-cold distilled water. Transfer the wet activated Sepharose into a 200 ml measuring cylinder and add the ice-cold lectin solution to be coupled (0.25 g of lectin dissolved in 0.5 M $NaCl-0.2$ M $NaHCO_3-0.2$ M haptenic sugar, pH 8.5). Stopper the measuring cylinder and gently stir for 24 h at 4°C. Filter the reaction mixture through a sintered glass filter funnel and wash the coupled gel extensively with 1 l each of ice-cold distilled water, 0.1 M $NaHCO_3$ and water, successively.

(iv) Block the unreacted iminocarbonate groups by suspending the preparation in 100 ml of a 1 M glycine solution, under gentle stirring, for 3 h at room temperature.

(v) Estimate the approximate amount of uncoupled lectin by measuring the absorbance at 280 nm (protein content) of the reaction mixture and recover the immobilized lectin by filtration, after a 24 h coupling period.

(vi) Wash the glycine-blocked gel extensively with cold distilled water and then with the equilibration buffer containing 0.02% sodium azide. Store the immobilized lectins at 4°C in this equilibration buffer until used.

### 4.2.2 *Immobilization of lectins on polyacrylhydrazido-sepharose (33)*

(i) To washed polyacrylhydrazido-Sepharose (Miles), add 3 vol. of 10% (w/v) glutaraldehyde solution, with gentle stirring, for 4 h at 4°C.

(ii)     Dissolve the lectin to be coupled in 0.1 M NaHCO$_3$ − 0.9 M NaCl − 0.1 M haptenic monosaccharide (4 − 5 mg of lectin/ml of coupling buffer) and add to the derivatized gel (4 − 5 mg of lectin/ml of settled gel beads). Stir the mixture slowly overnight.

(iii)    Wash the gel with phosphate-buffered saline (PBS) and add NaBH$_4$ dissolved in a gel volume of PBS (0.5 mg NaBH$_4$/ml of settled gel), for 3 h at 4°C. After washing with PBS containing 0.02% sodium azide, store the immobilized lectin at 4°C.

## 4.3 Fractionation of glycoproteins

### 4.3.1 *General remarks*

Affinity chromatography on different immobilized lectins has been widely used to purify various glycoproteins and many examples have been reviewed elsewhere (34,35) so that we shall restrict ourselves to some particular points which could be useful during such fractionations.

(i) *Use of crossed immuno-affinity electrophoresis of glycoproteins as a guide for lectin affinity chromatography:* Crossed immuno-affino electrophoresis (CIAE), introduced by Bøg-Hansen (36), combines the interaction between lectins and glycoproteins in the first dimension electrophoretic step with electrophoresis into an antibody-containing gel in the second dimension. This sensitive and powerful technique can give important information: (a) about interaction between a given lectin and the glycoprotein to be purified, a good correlation being generally observed between the results obtained by affinity electrophoresis and by affinity chromatography on the immobilized lectin; and (b) about the microheterogeneity of carbohydrate moieties of the glycoproteins to be purified. However, parameters such as column binding capacity and elution conditions cannot be predicted from the affinity electrophoresis experiments.

(ii) *Binding capacity of the immobilized lectins and problems of the elution:* Variations in the amount, as well as in the quality of the lectin immobilized on a gel, is an important factor influencing the binding of glycoproteins. Affinity differences are often observed between different commercially available immobilized Con A preparations. Similarly, affinity of WGA-Sepharose 4B for glycoproteins varies with the density of lectin coupled to the gel. This implies the calibration of the column before using a new batch of immobilized lectin with well known glycoproteins or glycopeptides such as $\alpha_1$-acid glycoprotein containing di-, tri- and tetra-antennary glycans (see *Figure 5.7*) and thyroglobulin containing glycans of the oligomannosidic type (see *Figure 4.1*). Three fractions are generally obtained, reflecting relative affinities of the immobilized lectins for the glycoproteins.

(a)     The *non-reactive compounds* are eluted at the void volume of the column with the equilibration buffer. This fraction, when the exact capacity of the immobilized lectin is not known, must be submitted to a new cycle of adsorption and elution on the same column, to be sure that the immobilized lectin was not saturated during the first run.

(b)     The *weakly reactive components* give a fraction which is obtained by elution with the equilibration buffer. The separation of weakly differently interacting glycoproteins can even be obtained by using a long and thin column which is more efficient than a wider column containing the same volume of immobilized lectin.

(c)     A *strongly reactive fraction* which is specifically desorbed by addition of the appropriate sugar at a concentration between 0.1 and 0.5 M. According to the commercial origin of the immobilized lectin, a weakly reactive glycoprotein can be either bound and eluted with the lectin-reactive fraction or unbound and eluted with the lectin-non-reactive fraction.

The spatial conformations of the native glycoprotein or non-specific hydrophobic interactions may modulate the accessibility of some glycans to the lectin, which can explain artefactual lectin-weakly reactive fractions. This inconvenience disappears when the glycoproteins are reduced and alkylated before fractionation on the immobilized lectin. In conclusion, the discovery of such lectin-weakly reactive glycoprotein fractions must be considered carefully.

Usually, immobilized lectin columns are eluted, after extensive washing with the equilibration buffer, with the same buffer but containing the haptenic monosaccharide (0.01 − 0.5 M). However, when glycoproteins differing in their lectin-reactive oligosaccharide residues are fractionated, it is interesting to elute the column with a concentration gradient of the haptenic sugar. Sometimes, the recovery of some bound glycoproteins is very low, even after elution with high concentration of haptenic sugar (0.5 M). This is the result either of a very high affinity of the lectin for some saccharidic determinants of the glycoproteins or of multivalent interactions between the immobilized lectin and a very high proportion of the saccharidic determinant on the glycoprotein. In the particular case of immobilized Con A, the recovery of high affinity glycoproteins can be improved by raising the temperature of the 0.5 M haptenic sugar solution to 37°C or even 60°C. Nevertheless, this procedure cannot generally be applied since most lectins are less efficient at 37°C than at room temperature (20°C) or 4°C.

The specific displacement of glycoprotein by haptenic sugar is reversible and after extensive washing with the equilibration buffer, the immobilized lectin can be used again with the same efficiency. However, stronger lectin−glycoprotein interactions can be displaced only with non-specific desorption processes which very often cause the irreversible denaturation of the immobilized lectin. Such non-specific displacements can be performed by pH change or with borate buffers of 0.02 − 0.1 M (pH 8.0), without denaturation of the lectin. High-yield recovery can also be obtained by heating the Con A − Sepharose − glycoprotein complex for 3 min at 100°C, in buffer containing 5% (w/v) sodium dodecyl sulphate and 8 M urea, but with irreversible inactivation of the lectin.

(iii) *Sequential affinity chromatography of glycoproteins on different immobilized lectins:* As is the case of oligosaccharides and glycopeptides (see below) the sequential use of chromatography on immobilized lectins, possessing different and well defined specificities toward saccharidic determinants, is used to fractionate complex mixtures of glycoproteins into classes, depending on their affinity for the different lectins. Immobilized Con A and WGA are the most currently utilized.

### 4.3.2 *Methods*

*Affinity chromatography of glycoproteins on immobilized Con A*

(i)    Equilibrate the lectin column (Con A – Sepharose 4B, 30 cm × 2.0 cm i.d.) in 5 mM sodium acetate buffer (pH 5.2) containing 0.1 M NaCl, 1 mM CaCl$_2$ and 1 mM MnCl$_2$ at a flow rate of 9 ml/h at room temperature. Use 0.02% sodium azide in all buffers.

(ii)   Dissolve the glycoprotein mixture (30 – 50 mg) in 3 ml of the same buffer and centrifuge in order to remove any insoluble material.

(iii)  Apply the clear glycoprotein solution to the column and elute with the equilibration buffer until the effluent is free from protein. Monitor the effluent absorbance at 280 nm and collect 2-ml fractions. In certain cases, Con A-weakly reactive variants can be recovered as a retarded fraction by elution with the equilibration buffer.

(iv)   Remove the weakly retained glycoproteins with 0.01 M methyl $\alpha$-D-glucoside in the equilibration buffer and the strongly Con A-reactive glycoproteins with 0.3 M methyl $\alpha$-D-glucoside solution.

(v)    Regenerate the Con A-column by extensive washing with the equilibration buffer.

(vi)   Pool the separated glycoprotein fractions and extensively dialyse against distilled water before freeze-drying. When the exact capacity of the Con A-column is not known, the unretained fraction must be submitted to a new fractionation cycle on the regenerated column.

*Affinity chromatography of glycoproteins on immobilized WGA*

(i)    Purify the WGA from wheat germ by affinity chromatography on a column of 2-acetamido-*N*-$\epsilon$-aminocaproyl-2-deoxy-$\beta$-glucopyranosylamine-Sepharose according to Lotan *et al.* (37). Couple the purified lectin to Sepharose 4B according to the procedure described above in order to obtain 5 mg immobilised WGA/ml of gel. Equilibrate the WGA-column (20 cm × 2.5 cm i.d.) in 0.01 M PBS (pH 7.2), containing 0.02% sodium azide, at a flow rate of 9 ml/h.

(ii)   Dissolve the glycoproteins (20 mg) in 3 ml of PBS and apply the clear glycoprotein solution to the column. Elute with PBS and monitor the effluent absorbance at 280 nm. As for Con A-affinity chromatography, WGA-weakly reactive glycoproteins can be recovered, as retarded fractions, by elution with the equilibration buffer.

(iii)  Elute the retained glycoproteins with 0.1 M *N*-acetylglucosamine in PBS. Pool the separated glycoprotein fractions and dialyse extensively against distilled water before freeze-drying.

(iv)   Regenerate the WGA column by extensive washing with PBS.

### 4.3.3 *Limitations of lectin affinity chromatography for purification of glycoproteins*

Affinity chromatography on immobilized lectins rarely succeeds in obtaining pure glycoproteins from complex mixtures. Generally, lectin affinity chromatography results in an enrichment of classes of different heterogeneous glycoproteins, possessing similar carbohydrate determinants recognized by the immobilized lectin and called 'lectin receptors'. As different lectins are able to recognize different saccharidic sequences belonging to the same glycan class and as these glycans are likely to be common to numerous glycoproteins, the different lectins will react in fact with a broad spectrum of glycoproteins.

In addition, non-specific interactions between glycoproteins and lectins often restrict the use of immobilized lectins for the fractionation of glycoproteins.

## 4.4 Fractionation of glycopeptides and oligosaccharides

If purification and fractionation of glycoproteins present some limitations which restrict the use of the affinity chromatography on immobilized lectins for isolating glycoproteins, the procedure represents, in contrast, a powerful tool for the fractionation of glycopeptides and glycans. This is essentially due to the fact that the above mentioned non-specific interactions often observed between glycoproteins and lectins are very rare in the case of glycans and glycopeptides so that these compounds interact specifically with the lectins and can be fractionated on the basis of an 'actual' affinity chromatography. Consequently, this implies, even more than in glycoprotein fractionations, a very precise knowledge of the exact specificities of the lectins to be used. That is to say that, once the precise specificity of an immobilized lectin is well defined, it becomes possible to predict the primary structure of bound glycopeptides or oligosaccharides.

### 4.4.1 *Affinity chromatography on immobilized Concanavalin A*

(i)     Equilibrate the Con A-Sepharose column (10 cm × 1.0 cm i.d.) in 5 mM sodium acetate buffer (pH 5.2), containing 0.1 M NaCl, 1 mM $CaCl_2$, 1 mM $MnCl_2$ and 1 mM $MgCl_2$, at a flow rate of 9 ml/h at room temprature. Use 0.02% sodium azide in all buffers.

(ii)    Dissolve the glycopeptides (or the oligosaccharides) in 1 ml of equilibration buffer and apply the clear solution to the column.

(iii)   Elute (1.5 ml fractions) successively with five column volumes of equilibration buffer, then with three column volumes of 0.01 M methyl α-D-glucoside and finally with five column volumes of 0.3 M methyl α-D-glucoside.

In this way, oligosaccharides and glycopeptides are fractionated into three classes (*Figure 13*).

(i)     *Non-reactive glycopeptides* (or *oligosaccharides*), eluted at the void volume of the column with the equilibration buffer.

(ii)    *Weakly reactive derivatives* obtained by eluting with the equilibration buffer itself. However, depending on the amount of immobilized Con A/ml of Sepharose, the same type of compounds can be weakly retained and may be eluted as a sharp peak with low concentration of the haptenic sugar derivative (0.01 M methyl α-D-glucoside).

(iii)   *Strongly reactive components*, eluted as a sharp or broad tailing peak with 0.3 M methyl α-D-glucoside, depending on the Con A concentration in the gel.

Purify the fractions from salts and haptenic sugar by gel filtration on a Bio-Gel P-2 column (50 cm × 2 cm i.d.) equilibrated with distilled water. After elution with the

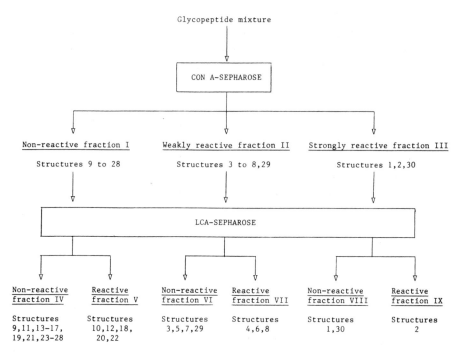

**Figure 13.** Scheme of fractionation of *N*-glycosylpeptides by combining affinity chromatography on immobilized ConA- and LCA-Sepharose. Arabic numerals refer to glycopeptide structures of *Table 3*.

last haptenic sugar solution, regenerate the Con A column by extensive washing with the equilibration buffer.

*General comments*

(a) The saccharidic structures (listed in *Table 3*) constituting each of the obtained fractions are described in *Figure 13*.

(b) Glycans released from glycosylpeptides, either by hydrazinolysis or by the action of various endo-*N*-acetyl-β-D-glucosaminidases, present the same behaviour on immobilized Con A as glycopeptides.

(c) Sometimes, high affinity oligosaccharides are eluted from the column only by raising the temperature of the sugar solution to 60°C or by irreversible denaturation of the lectin with two column volumes of a 1% (w/v) solution of SDS. In such a case, the column cannot be re-used.

(d) Before using a new batch of immobilized Con A, calibration of the immobilized lectin with well-defined glycopeptides must be performed.

(e) Extensive use leads to a decrease of the binding capacity of Con A-Sepharose due to lectin leakage.

(f) When the exact binding capacity of the gel is not known, the unretained fraction must be recycled using the regenerated column. Commercially available Con A-Sepharose (Pharmacia) is able to bind 50−75 μg of biantennary *N*-glycosylpeptides, obtained from human serotransferrin, per ml of gel.

(g) Before chromatography, glycopeptides can be [14]C- or [3]H-labelled by *N*-acetylation according to the general procedure described above (see Section 2.2.2), by using labelled acetic anhydride. In the same way, oligosaccharides can be labelled by reducing the terminal monosaccharide residue with tritiated sodium borohydride by applying the general procedure of reduction and purification as described in Section 2.1.2.

### 4.4.2 *Affinity chromatography on immobilized Lens culinaris agglutinin*

LCA-Sepharose 4B can be purchased from different manufacturers or easily prepared from *Lens culinaris* seeds (38) and coupled to CNBr-activated Sepharose 4B, at a concentration of 5 mg of lectin/ml of gel.

(i)   Equilibrate the lectin column (15 cm × 2.5 cm i.d.) in PBS (pH 7.4) containing 0.02% sodium azide (divalent cations such as $Ca^{2+}$ or $Mn^{2+}$ are not required), at a flow rate of 9 ml/h at room temperature.

(ii)  Dissolve the glycopeptides in 1 ml of PBS and apply the clear solution to the column.

(iii) Elute (1.5 ml fractions) the non-reactive glycopeptides (at the void volume) by passing five column volumes of PBS, and the lectin-reactive glycopeptides with five column volumes of PBS containing 0.15 M methyl $\alpha$-D-glucoside. As for immobilized Con A, fractionation can be easily followed by labelling the glycopeptides by *N*-acetylation with [14C]- or [3H]acetic anhydride.

(iv)  Purify the fractions from salts and haptenic sugar by gel filtration on a Bio-Gel P-2 column (50 cm × 2 cm i.d.) equilibrated with distilled water.

(v)   Regenerate the LCA column by extensive washing with PBS.

The following comments apply to this procedure:

(a)   The saccharidic structures consisting each of the fractions are described in *Figure 13*. From these results, it clearly appears that the $\alpha$-1,6-linked fucose residue is determinant for the binding of glycopeptides and that affinity chromatography on immobilized LCA is a valuable tool for separating glycopeptides into two classes: $\alpha$-1,6-fucosylated and non-fucosylated.

(b)   Only glycopeptides can be fractionated on immobilized LCA, not oligosaccharides. So, glycans released from *N*-glycosylpeptides by hydrazinolysis/re-*N*-acetylation or from *O*-glycopeptides by reductive $\beta$-elimination do not interact with immobilized LCA and are eluted in the void volume.

(c)   As in the above case, immobilized *Vicia faba* and *Pisum sativum* lectins interact only with fucosylated glycopeptides. However, their affinity is weaker, so that the fractionation into the two classes is achieved by using only the equilibration buffer.

### 4.4.3 *Affinity chromatography on other lectins*

Other immobilized lectins are also used, but to a lesser extent, for fractionating glycopeptides or oligosaccharides: PHA, WGA, RCA and DSA. In all cases, the general procedures are identical to that described for immobilized Con A and LCA. They differ only by the method of elution (*Table 4*).

The following general comments apply to this procedure:

**Table 4.** Composition of the equilibration and elution buffers and of the lectin-reactive fractions obtained.

| Immobilized lectin | Mode of elution | | Composition of the lectin-reactive fractions[a] |
| --- | --- | --- | --- |
| | Equilibration buffer | Elution buffer | |
| L$_4$-PHA[b] | PBS | PBS | 17, 18, 21, 22 |
| E$_4$-PHA[c] | PBS | PBS | 9, 10, 27, 28 |
| WGA | PBS | 0.1 M GlcNAc in PBS | 11, 29 |
| RCA$_1$ | PBS | 0.15 M Gal in PBS | 5, 6, 13, 14, 17, 18, 21, 22 |
| DSA | PBS | 8 mg/ml of a mixture of *N,N'*-diacetylchitobiose and *N,N',N''*-triacetylchitobiose in PBS | 25, 26 |

[a]Numbers refer to the structures described in *Table 3*.
[b]*Phaseolus vulgaris* leukoagglutinating lectin.
[c]*Phaseolus vulgaris* erythroagglutinating lectin.

(a)     *N,N'*-diacetylchitobiose-Asn inner-core is an important determinant for binding to immobilized WGA and LCA, so that free glycans and oligosaccharides, devoid of the Asn residue, do not interact with the immobilized lectins. Substitution by an α-L-fucose residue at the C-6 position of the *N*-acetylglucosamine residue involved in the *N*-glycosylamine bond reduces the affinity of immobilized WGA for glycopeptides.

(b)     Immobilized WGA can also interact with glycopeptides possessing numerous clustered *O*-glycosidically linked sialyloligosaccharides.

(c)     Asialo, bi-, tri- and tetra-antennary glycans of the *N*-acetyllactosamine type can be fractionated on immobilized RCA$_I$. However, the separation depends upon the amount of bound RCA$_I$/ml of gel so that immobilized RCA$_I$ columns must be carefully calibrated with known glycopeptides before use.

(d)     Some lectins possess a specificity directed toward the terminal non-reducing sugars and are interesting tools for the fractionation of particular glycopeptides. For example, the immobilized A$_4$ tetrameric isolectin from *Griffonia simplicifolia* agglutinin I (GSA$_I$-A$_4$) presents a strict specificity toward *N*-acetyl-α-D-galactos-amine in the terminal non-reducing position (39), while the immobilized B$_4$ tetrameric isolectin from *Griffonia simplicifolia* agglutinin I (GSA$_I$-B$_4$) is specific for glycopeptides with terminal non-reducing α-D-galactose residues (40). Another lectin (GSA$_{II}$), also isolated from the seeds of *Griffonia simplicifolia*, is a very useful tool for fractionating glycopeptides or oligosaccharides with *N*-acetyl-α-D-glucosamine residues in a terminal non-reducing position (41).

(e)     Two immobilized fucose-binding lectins [*Lotus tetragonolobus* (LTA) and *Ulex europeus* (UEA$_I$) agglutinins] have no affinity for fucosylated glycopeptides or oligosaccharides of the *N*-acetyllactosamine type. Only the Fuc(α1-6)GlcNAc(β1-N)Asn glycopeptide is retained on both immobilized lectins and is eluted with 0.1 M L-fucose in PBS. However, a new fucose specific lectin, isolated from the mushroom *Aleuria aurantia*, seems to be a very promising tool for fractionating fucosylated glycopeptides and oligosaccharides (42).

### 4.4.4. *Sequential affinity chromatography on immobilized lectins*

Most of the *N*-glycosylpeptides can be fractionated into relatively homogeneous classes by sequential affinity chromatography (28,43). Up to now, the associations Con A- and LCA—Sepharose and of Con A-, LCA- and L$_4$-PHA—Sepharose are the most commonly utilized for fractionating the glycopeptides obtained from cell membrane glycopro-

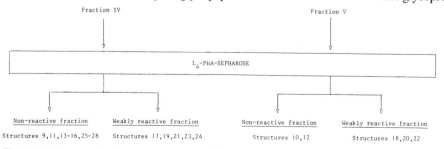

**Figure 14.** Scheme of subfractionation on immobilized L$_4$-PHA of *N*-glycosylpeptide fractions IV and V from *Figure 13*. Arabic numerals refer to glycopeptide structures of *Table 3*.

teins. This methodology is rapid and sensitive, particularly when glycopeptides are radiolabelled by using [14]C- and/or [3]H-labelled precursors or by *N*-acetylation carried out with [[14]C]acetic anhydride. *Figure 13* presents the six classes of glycopeptides obtained by sequential affinity chromatography on immobilized Con A- and LCA−Sepharose and *Figure 14* shows the results obtained by subfractionating fractions IV and V isolated by this procedure on immobilized *Phaseolus vulgaris* leukoagglutinating lectin $L_4$-PHA.

## 5. COLORIMETRIC ASSAYS OF CARBOHYDRATES IN GLYCOPROTEINS AND GLYCOPEPTIDES

### 5.1 Glycan composition

The composition of a glycoprotein or of a glycan in terms of neutral sugars, hexuronic acids, hexosamines and sialic acids is usually determined by specific colorimetric procedures which have been described in Chapter 1 (see also ref. 15 − 18).

The determination of neutral monosaccharides, of hexuronic acids and, in some cases, of sialic acids is carried out without prior hydrolysis of the glycosidic linkages since the conditions of the reactions developed in presence of sulphuric acid ensure the complete breakdown of all glycosidic bonds. This is not the case for the determination of hexosamines which requires the prior liberation of the amino sugars, as well as for the determination of sialic acids by the thiobarbituric acid method. Therefore, one of the major problems concerns the risk of destruction of neutral monosaccharides and hexuronic acids during the reaction in acidic medium, on the one hand, and of incomplete liberation of hexosamines and sialic acids by chemical or enzymatic hydrolysis, on the other hand. In order to limit and to control the destruction of neutral monosaccharides and hexuronic acids, the temperature of reaction mixtures must be identical in all the test tubes during the assay and the rate of flow of the acid down the test tube, the mixing of solutions and the cooling must be standardized. In order to obtain maximal liberation of conjugated hexosamines and sialic acids without degradation of the liberated sugars, kinetics of hydrolysis using increasing concentrations of acidic or enzymatic agent have to be determined on each sample of glycan or glycoprotein.

The second major problem encountered during sugar determinations concerns the specificity of the reaction. In fact, the determination of neutral sugars and hexuronic acids is based on the production of furan derivatives in hot acid solution followed by the condensation of the chromogens with a specific reagent. The formed chromophores develop a characteristic coloration which is proportional to the sugar concentration. However, the specific absorbances show a wide variation between the monosaccharides. In order to eliminate this cause of error, the internal standards must be prepared with a mixture of monosaccharides in the same molar ratio as expected in the analysed compound. In each series of determinations, appropriate blanks must be used containing the solution used for dissolving the analysed samples.

### 5.2 Colorimetric determination of neutral monosaccharides

Solutions of phenol, cysteine and orcinol in sulphuric acid are the reagents most commonly used for quantitative determination of hexoses, 6-deoxyhexoses, methylpentoses and pentoses (see Chapter 1, Sections 2.1 and 2.3).

## 5.3 Hexuronic acids

Naphtoresorcinol, sulfhydryl compounds and carbazole in hydrochloric or sulphuric acid solutions are the most commonly reagents used for the detection and the quantitative determination of hexuronic acids. We shall restrict ourselves to the description of the methods of Bitter and Muir's using carbazole reagent (see Chapter 1, Section 2.6) and Blumenkrantz and Asboe-Hansen using the *meta*-hydroxydiphenyl. This latter method is recommended for determining the acid mucopolysaccharides in biological materials.

### 5.3.1 *Meta-hydroxydiphenyl-sulphuric assay*

*Reagents:*

(A)   0.15% (w/v) solution of *m*-hydroxydiphenyl in 0.5% (w/v) NaOH. The reagent is stable at 4°C for more than 1 month.
(B)   0.0125 M solution of sodium tetraborate in concentrated sulphuric acid.
(C)   Standard aqueous solution containing 100 $\mu$g of uronic acids/ml.

### *Method*

(i)    Add 1.2 ml of sulphuric-tetraborate to 0.2 ml of glycan, glycoprotein (0.5−20 $\mu$g of total uronic acids) or standard solution.
(ii)   Cool the tubes in crushed ice, shake vigorously.
(iii)  Maintain the tubes in a boiling water-bath for 5 min, cool in a water-ice bath and add 20 $\mu$l of the m-hydroxydiphenyl reagent.
(iv)   Shake the tubes and determine the absorbances at 520 nm.

The absorbances given by glucuronic, iduronic and galacturonic acids are very similar, while the absorbance given by mannuronic acid is about half this value. The interference of hexoses is very low. Therefore, the method presents the advantage of the specificity.

## 5.4 Hexosamines and N-acetylhexosamines

All the procedures for the determination of free hexosamines derive from the Elson-Morgan reaction which involves a condensation with acetylacetone in alkaline solution leading to the formation of pyrrole derivatives which develop with the *p*-dimethylamino-benzaldehyde Ehrlich's reagent a specific pink coloration. The method proposed by Belcher *et al.* (71) will be described here.

Free *N*-acetylhexosamines are determined according to a different procedure proposed by Morgan and Elson and based on pink chromophore production by direct heating with the Ehrlich's reagent in alkaline solution (see Chapter 1, Section 2.5). On the same principle, free hexosamines can be determined after re-*N*-acetylation.

### 5.4.1 *Colorimetric determination of hexosamines (71)*

*Hydrolysis.* Hydrolysis of the very stable hexosaminyl linkages is usually carried out with 4 M HCl at 100°C for 4 h. However, as the stability of this type of linkage depends on the structure of the glycans, this procedure cannot be considered as a general one. The best way is to determine for each glycoprotein, the optimal hydrolysis conditions by using HCl at concentrations ranging from 2 to 6 N and by determining the kinetics of hydrolysis (2 – 10 h). HCl is then removed under vacuum in presence of NaOH. The dry product is dissolved in water and centrifuged.

*Reagents*

(A)   Acetylacetone reagent freshly prepared by adding 2 ml of acetylacetone to 48 ml of 0.625 M $Na_2CO_3$ (dried overnight at 100°C).

(B)   Ehrlich's reagent prepared by dissolving 1.6 g of recrystallized *p*-dimethylamino-benzaldehyde in 30 ml of concentrated HCl.

(C)   Standard aqueous solution containing 100 $\mu$g of free hexosamines/ml.

*Method*

(i)    Add 0.5 ml of acetylacetone reagent and 1 ml of distilled water to 0.5 ml of glycan, glycoprotein (5–50 $\mu$g of total hexosamines) or standard solution.

(ii)   Shake the tubes vigorously, cap with glass marbles to minimize evaporation and heat in a boiling water-bath for 10 min.

(iii)  Cool to room temperature, add 2.5 ml of ethanol, mix carefully and maintain the tubes in a water-bath at 75° ± 2°C for 5 min.

(iv)  Add 0.5 ml of Ehrlich's reagent and heat at 75°C for 30 min.

(v)   Cool the tubes to room temperature and add 2.5 ml of 95% ethanol.

(vi)  Determine the absorbance at 520 nm after 30 min. The colour is stable for 24 h.

  Some important notes concerning this method are:

(i)    The absorption curve of hexosamines shows a maximum at 520 nm. As the relative extinction coefficients are different (glucosamine = 100; galactosamine = 140), standard solutions must contain a mixture of glucosamine and galac-tosamine in the same ratio as in the analysed sample.

(ii)   Muramic acid and *N*-acetylhexosamines interfere. Sialic acids react by produc-ing a blue coloration. Hexoses and uronic acids do not interfere.

(iii)  Sodium tetraborate and sodium chloride decrease the intensity of the coloration. If the hydrochloric acid used to liberate the hexosamines is neutralized by NaOH, the standard solution of hexosamines has to be prepared in a NaCl solution hav-ing the same concentration as that of neutralized hydrolysates.

## 5.5 Sialic acids

Sialic acids are determined by two different types of procedures. The first group of methods leads to the determination of free and conjugated sialic acids and is based on the reaction of chromogens obtained in acidic medium with reagents such as resor-cinol, *p*-dimethylaminobenzaldehyde and diphenylamine. The second group of methods is applicable only to free sialic acids.

### 5.5.1 Determination of free and conjugated sialic acid (72)

*Reagents*

(A)   Diphenylamine reagent (Dische's reagent) prepared by dissolving 1 g of diphenyl-amine (recrystallized from ethanol) in a mixture of 90 ml of glacial acetic acid and 10 ml of concentrated sulphuric acid. Store the reagent at 4°C in the dark and transfer to a water bath at 40°C just before use in order to re-dissolve the crystals.

(B)     Solution of 7.5% (w/v) trichloroacetic acid in water.

(C)     Standard aqueous solution containing 100 $\mu$g of sialic acid/ml.

*Method*

(i)     Add 0.2 ml of trichloroacetic acid at 7.5% and 6 ml of diphenylamine reagent to 0.1 ml of glycan, glycoprotein (1 − 20 $\mu$g of sialic acid) or standard solution.

(ii)    Mix and heat the tubes in a boiling water-bath for 30 min.

(iii)   Cool the tubes in water and determine the absorbance of the violet-blue coloration at 530 nm after a 30 min stay in the dark. Coloration is stable for 6 h.

Some important notes concerning this method are:

(i)     The extinction coefficient of *N*-glycolylneuraminic acid at the $\lambda_{max}$ 530 nm is 83% of that of *N*-acetylneuraminic acid. Therefore, it is important to use standards containing the two sialic acids in the same ratio as that expected in the analysed solution.

(ii)    At high concentrations, hexoses yield a blue colour with absorption maxima at 530 nm and 650 nm, as well as ketohexoses whose absorption curve presents maxima at 530 nm and 640 nm. In both cases, molar absorption of neutral sugars and *N*-acetylneuraminic acid are identical. 2-Deoxypentoses, particularly 2-deoxyribose, produce a blue colour with a characteristic spectrum with an absorption maximum at 610 nm. In addition, the molar absorption is twice as high as that of *N*-acetylneuraminic acid. Therefore, the complete absorption spectrum must be determined in order to avoid interference due to the presence of DNA. Methylpentoses, hexosamines and *N*-acetylhexosamines do not interfere in the reaction. Glycerol produces a green colour.

### 5.5.2 *Quantitative determination of free sialic acids*

Due to the external position of sialic acids in glycans and to the lability of their glycosidic linkages, these sugars can be easily removed under very mild acidic conditions or by using neuraminidases. Oxidation of free *N*-acetyl or *N*-glycolylneuraminic acids with periodic acid leads to the formation of formyl-pyruvic acid which reacts with 2-thiobarbituric acid and produces a pink colour with an absorption maximum at 549 nm (see Chapter 1, Section 2.7). Details of the colorimetric assays of *N,O*-acylneuraminic acids are described in Reference 19.

### 5.5.3 *Hydrolysis of sialyl linkages*

*Chemical hydrolysis.* Liberation of sialic acids is usually carried out with 0.05 M sulphuric acid at 80°C for 1 h or with 0.05 M sulphuric acid at 100°C for 1 h. Under these conditions, all types of sialic acids are removed without any degradation, except *O*-deacetylation. Moreover, the other monosaccharides, with the exception of fucose, are not liberated. Hydrochloric acid is not recommended since lactonisation and partial destruction of sialic acids have been observed.

*Enzymatic hydrolysis.* Enzymatic hydrolysis of sialic acid linkages can be achieved by use of neuraminidases from *Clostridium perfringens, Vibrio cholerae* or Influenza virus. The specificity of these enzymes as well as the hydrolysis conditions are described in Section 8.2.1.

## 6. IDENTIFICATION AND DETERMINATION OF GLYCAN MONOSAC-CHARIDES BY GAS-LIQUID CHROMATOGRAPHY

### 6.1 Glycoprotein hydrolysis

For a long time, the determination of the molar composition of glycans encountered an important problem: that of the quantitative liberation of monosaccharides which must be achieved before any chromatographic analysis. In fact, all the glycosidic bonds must be split without any destruction of the liberated sugars. In addition, these bonds vary in stability depending on the nature of the monosaccharides and on the type of glycosidic linkage; the sialyl bonds being the most labile and the uronosidyl and glucosaminyl ones being the most stable toward hydrochloric and sulphuric acids. Therefore, the use of these acids necessitates the determination, for each type of sugar (i.e. hexoses, methylpentoses, hexosamines, uronic acids and sialic acids), of different conditions of acid molarity, time and temperature of hydrolysis. A compromise between the liberation of the monosaccharides and their destruction is obtained. The failure of aqueous hydrochloric or sulphuric acid hydrolysis has been overcome by the introduction, on the one hand, of hydrolysis by trifluoroacetic acid even if sialic acids are destroyed and, on the other hand, of methanolysis in which anhydrous methanol−HCl mixtures are used. All kinds of monosaccharides, including sialic acids, are stable to methanolysis. In any case, the gas liquid chromatography is the best method for identification and determination of liberated monosaccharides (for review, see refs. 44 and 45).

### 6.2 The use of alditol acetates

G.l.c. of alditol acetates is widely used for determining the composition of monosaccharide mixtures. In fact, the main difficulty encountered in the separation of the monosaccharides themselves is the formation of at least four glycosides per monosaccharide resulting from anomeric and ring isomerization, each of which giving a peak on the chromatogram. Thus, in a complex mixture containing several monosaccharides, the multiplicity of peaks leads to overlappings so that accurate determinations cannot be achieved (see also Chapter 1, Section 6). On the contrary, an alditol acetate originating from one monosaccharide gives rise to only one peak.

### 6.2.1 *Hydrolysis of glycoproteins*

Hydrolyse amounts of glycoproteins corresponding to 10 $\mu$g of total sugars with 100 $\mu$l of 4 M trifluoroacetic acid in the presence of 2 $\mu$g of mesoinositol used as internal standard, at 100°C for 4 h, under helium, in glass tubes fitted with teflon-lined screw caps (Sovirel, Paris, France; also Pierce). After cooling, evaporate the solution in a dessicator, under vacuum. Do not forget that, in these conditions, sialic acids are total-ly destroyed.

## 6.2.2 *Reduction and analysis of monosaccharides*

This is described in Chapter 1, Section 6.2.

## 6.3 **The use of methylglycoside trifluoroacetates**

In a first step, the *O*-glycosidic bonds are quantitatively split by methanolysis, leading to the formation of *O*-methylglycosides which are trifluoroacetylated in a second step. The trifluoroacetates of *O*-methylglycosides (TFA-derivatives) are easily separated by g.l.c. with little overlapping. Sialic acids and uronic acids are stable toward methanolysis. We currently use a method derived from the procedure described by Zanetta *et al.* (for review, see ref. 44).

(i)      The methanolysis reagent is prepared as detailed in Chapter 1, Section 6.1.

(ii)     Freeze-dry carefully (complete dehydration of samples is the main condition of success) amounts of glycoproteins corresponding to 10 $\mu$g of total sugars to which 2 $\mu$g of mesoinositol are added as internal standard.

(iii)    Treat the residue with 200 $\mu$l of 0.5 M methanol$-$HCl for 24 h at 80°C in glass tubes fitted with teflon-lined screw caps.

(iv)    Cool, remove HCl and methanol under a stream of nitrogen at 50°C.

(v)    Trifluoroacetylate by dissolving the dried residue in a mixture of dichloromethane (50 $\mu$l) and trifluoroacetic anhydride (50 $\mu$l). Stopper the tube quickly and keep at room temperature overnight.

(vi)    G.l.c. of TFA-*O*-methylglycosides is carried out on 10 $\mu$l samples by using a

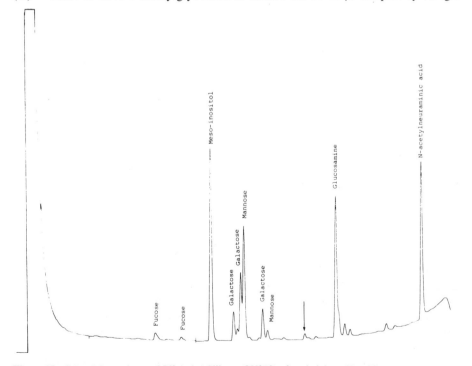

**Figure 15.** G.l.c. (glass column of 5% (w/w) Silicone OV210) of methylglycoside trifluoroacetates obtained from $\alpha_1$-acid glycoprotein. Arrow indicates the trifluoroacetate derivative of glucosamine.

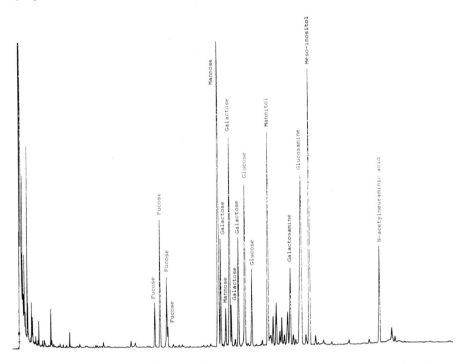

**Figure 16.** G.l.c. (glass capillary column of Silicone OV101) of trimethylsilylated methylglycosides of fucose, mannose, galactose, glucose, galactosamine, glucosamine and *N*-acetylneuraminic acid.

glass column (300 cm × 0.3 cm i.d.) filled with 5% (w/w) Silicone OV 210 on Chromosorb W-HP (mesh 120 – 140) and a Varian Aerograph model 1400 chromatograph equipped with a flame ionization detector. The carrier gas is helium at a flow rate of 7.5 ml/min. The oven temperature is programmed from 110°C to 220°C at 2°C/min with injector and detector temperatures of 225°C and 250°C, respectively. A typical gas liquid chromatogram of the methylglycoside trifluoroacetates obtained from $\alpha_1$-acid glycoprotein (orosomucoid) is presented in *Figure 15*.

## 6.4 The use of trimethylsilylated methylglycosides

This is described in Chapter 1, Section 6.1. Before carbohydrate analysis, purify all glycoproteins (10 μg) by rapid gel-filtration on Aquagel type P-2 10 μm (Polymer Laboratories, Shropshire, England) column (30 cm × 0.7 cm i.d.) fitted to a Spectra-Physics Model 8770 liquid chromatograph (with a 500 μl sample loop), eluting with water (0.4 ml/min). Detect glycoproteins with an u.v.-detector (L.K.B. 2138 Uvicord S) at 206 nm. Collect the excluded fraction in a glass tube fitted with teflon-lined screw cap and freeze-dry in presence of 0.1 μg of internal standard mannitol. A typical gas-liquid chromatogram is given in *Figure 16*.

### 6.4.1 *General comments on the g.l.c. of glycoprotein derived monosaccharides*

(i)     Methanolysis offers, as compared to hydrolysis, the considerable advantage of cleaving all *O*-glycosidic linkages in a one-step procedure. However, we must keep in mind the following failures:

(a) Under the usual conditions (0.5 M methanolic HCl, 24 h, 80°C), the linkage between GlcNAc and Asn is split to only a very limited extent and the glucosamine is liberated as a free monosaccharide (see *Figure 15*) and not as a methylglycoside.

(b) Dehydration of monosaccharides can occur. For example, *N*-acetyl-neuraminic acid gives 3% of 2,7 anhydro-*N*-acetylneuraminic acid, and alditols, like 2-acetamido-2-deoxy-D-galactitol coming from glycan removed by β-elimination in reductive conditions (see paragraph 2.1), also give anhydro-derivatives.

(ii) Methanolysis must be carried out under anhydrous conditions. Thus, (a) carefully dry all materials and samples; and (b) be careful not to introduce water in the methanol-HCl reagent!

(iii) The lower limit of sensitivity is not due to the detector response but to contaminants which accumulate during the operations (e.g. glucose and xylose released from dialysis membranes, glucose from the Sephadex columns or from sucrose gradients used for fractionation of glycoproteins, and various materials from insufficiently cleaned glassware).

## 7. METHYLATION

Methylation analysis is one of the most powerful procedures for structural analysis of polysaccharides and glycoprotein glycans (for reviews, see ref. 45 – 47). It involves the permethylation of all free hydroxyl groups of the sugars followed by the liberation of methylated monosaccharides by hydrolysis or methanolysis and the qualitative and quantitative analysis of the methylated derivatives. The positions of the free hydroxyl groups of the partially methylated monosaccharides gives the positions in which the sugar residues was glycosylated. Many methylation methods have been reported since the first procedure described in 1903 by Purdie and Irvine but most of the proposed methods led to an incomplete methylation of saccharides. In this connection, Hakomori's procedure (48) using sodium hydride and dimethylsulphoxide represented real progress by producing a more powerful nucleophile than the bases previously used. The method leads to a rapid and complete *O*-methylation and *N*-methylation (of the acetamido group of hexosamine residues) without any loss of *N*-acetyl groups.

Because the sodium hydride contains impurities (mainly due to the oil in which it is dispersed), cleaner reagents were proposed to perform the methylation of small amounts of sugars. In this connection, we have produced a procedure based on the use of dimethylsulfoxide and *n*-butyl lithium which gives less background in g.l.c. and, thus, can be used to methylate micro-quantities of glycoprotein glycans ( < 10 μg total sugars) (49).

### 7.1 The use of potassium tert-butoxide in dimethylsulphoxide

#### 7.1.1 *Preparation of potassium tert-butoxide*

(i) Treat 5 g of potassium with 25 ml of dry *n*-pentane in an atmosphere of helium and add 25 ml of freshly distilled *tert*-butyl alcohol until the evolution of hydrogen starts. Repeat the washing twice (yield: 4.7 g of clean potassium).

(ii)   Place the potassium (4.7 g), under a stream of helium, in a one necked flask containing 100 ml of *tert*-butyl alcohol and equipped with a reflux condenser connected to a drying tube packed with calcium chloride.

(iii)  Heat the mixture for 30 min at $40 - 50°C$ using an electric heating mantle. Reflux the mixture for $2 - 3$ h increasing the temperature gradually to $80 - 81°C$. When the reaction is complete cool the flask rapidly to room temperature.

(iv)   Evaporate the *tert*-butyl alcohol at $40°C$ in a carefully washed and dried rotary evaporator. During the evaporation, the receiver flask of rotary evaporator is cooled in an ice−acetone bath. Eliminate the remaining *tert*-butyl alcohol by lyophilisation in a helium atmosphere. The yield of potassium *tert*-butoxide is 14 g.

### 7.1.2 *Preparation of potassium methyl-sulphinyl carbanion*

Add rapidly 14 g of potassium *tert*-butoxide in 42 ml of dry and freshly distilled dimethylsulphoxide (DMSO) and place in an ultrasonic bath for 30 min. The base concentration of the reagent is 3 M. Before using, heat at $60°C$ for 30 min.

### 7.1.3 *Micro-methylation procedure*

Methylation is carried out in small glass tubes ($6 \times 80$ mm) placed in teflon-lined screwcap tubes ($13 \times 100$ mm).

(i)    Dissolve 10 $\mu g$ of glycopeptides or reduced oligosaccharides in 50 $\mu l$ of DMSO and 50 $\mu l$ of potassium methylsulphinyl carbanion under an inert atmosphere.

(ii)   Sonicate the mixture for 60 min. After cooling at $-4°C$, add 100 $\mu l$ cold methyl iodide.

(iii)  Sonicate the solution at $20°C$ for 45 min. An important point is to alkylate at room temperature to avoid the decomposition of methyl iodide to iodine.

(iv)   Stop the methylation by the addition of 0.5 ml water and of a few crystals of sodium thiosulfate.

(v)    Extract the permethylglycan once with 0.3 ml of chloroform and twice with 0.5 ml of chloroform. Wash the chloroform phase (1.3 ml) five times with 0.5 ml water.

(vi)   Dry the chloroform extract by the addition of anhydrous sodium sulphate, filter and concentrate in a rotary evaporator. Eliminate the last traces of DMSO by lyophilisation.

## 7.2 **The use of lithium methylsulphinyl carbanion**

### 7.2.1 *Preparation of lithium methylsulphinyl carbanion*

The apparatus used for preparing the lithium methylsulphinyl carbanion consists of a 500 ml two-necked flask provided with a magnetic stirrer, fitted with a 250 ml dropping funnel and with a distillation system equipped with a drying tube packed with calcium chloride, a manometer and an one-necked 1 litre trapflask. The apparatus is connected to a water-pump. Before the reaction procedure, the system is purged using a helium stream.

(i)     Introduce 120 ml of freshly dried and distilled DMSO (Merck, Darmstadt, FRG) into the two-necked flask under a helium atmosphere. Place a solution of *n*-butyl lithium (Janssen Chimica, Beerse, Belgium; 1.6 M) in hexane (250 ml) in the dropping funnel.

(ii)    Add this solution gradually at room temperature, under reduced pressure (30 mm) with constant stirring.

(iii)   Remove the *n*-butane (produced during the reaction of *n*-butyl lithium with DMSO) and hexane under vacuum. The addition is carried out in the course of about 2 h after which the colour of the reaction mixture changes to dark green. After complete addition of the *n*-butyl lithium solution, maintain the mixture at low pressure for 2 h at room temperature.

(iv)   Bring the system back to atmospheric pressure with the admission of helium. Store the reagent at 4°C under helium atmosphere in teflon-lined screw-cap tubes (13 × 100 mm).

### 7.2.2 *Micromethylation procedure:*

(i)    Dissolve 10 μg of reduced oligosaccharides in 20 μl of DMSO in small glass tubes (6 × 80 mm) placed in teflon-lined screw-cap tubes (13 × 10 mm).

(ii)    Add 20 μl of lithium methylsulphinyl carbanion under an inert atmosphere and sonicate the mixture at 20°C for 60 min.

(iii)   Cool to −4°C, add cold methyl iodide (40 μl) and sonicate the solution at 20°C for 45 min.

(iv)   Stop the methylation by the addition of water (0.4 ml) and sodium thiosulphate. Extract the permethylated oligosaccharide-alditols three times with 0.4 ml of chloroform.

(v)    Wash five times with 0.5 ml of water, filter, concentrate and evaporate the chloroform phase at 25°C.

## 7.3 Analysis of monosaccharide methyl ethers

The combination of gas-liquid chromatography with mass spectrometry allows an easy identification of methylated monosaccharides obtained by hydrolysis or methanolysis of permethylglycans. In addition, chemical ionization mass spectrometry greatly enhances the sensitivity of detection of methylated sugars by removing the background noise from the gas chromatograms.

### 7.3.1 *Methanolysis of permethylglycans*

The methylated sugars are analysed as their partially methylated and acetylated methyl-glycosides as follows:

(i)    Treat the permethyloligosaccharide-alditols (1 − 10 μg) with 0.5 M methanol-HCl (150 μl) in a small glass tube (6 × 80 mm) placed in teflon-lined screw-cap tube (13 × 100 mm) for 20 h at 80°C.

(ii)    Eliminate HCl−methanol reagent under a stream of nitrogen.

(iii)   Acetylate the methylated methylglycosides in the same tube with 40 μl pyridine−acetic anhydride (1/1 v/v) at 37°C for 24 h.

(iv)   Evaporate the reagent under a stream of nitrogen.

(v)    Dissolve the methylated acetylated methylglycosides in methanol and inject into
       a gas chromatograph coupled to a mass spectrometer.

### 7.3.2 Gas-liquid chromatography-mass spectrometry

The methylglycoside methyl ethers are analysed by g.l.c.-m.s. under the following experimental conditions in our laboratory: Girdel model 300 g.l.c. apparatus (Suresnes,
France); glass capillary column (60 m × 0.35 mm i.d.) coated with Silicone OV 101;
helium pressure: 0.4 bar; column temperature: $110-240°C$ at $3°C/min$. The retention times of the methylglycoside methyl ethers are calculated relative to the palmityl
methyl ester. Mass spectra are recorded on a Riber-Mag 10-10 mass spectrometer (Rueil-
Malmaison, France) using an electron energy of 70 eV and an ionizing current of 0.2
mA for electron impact (e.i.) mode. Retention times of methylated and acetylated

**Table 5.** Retention time of $O$-acetyl-$O$-methyl derivatives of methyl 2-deoxy-2-($N$-methylacetamido)D-
glucoside relative to palmityl methyl ester.

| Position of substituent | | Retention times (min) | |
|---|---|---|---|
| | | Peaks | |
| Methyl | Acetyl | 1 | 2 |
| 3,4,6 | | 0.80 | 0.96 |
| 4,6 | 3 | 1.06 | 1.08 |
| 3,6 | 4 | 1.03 | 1.15 |
| 3,4 | 6 | 1.11 | 1.21 |
| 6 | 3,4 | 1.18 | 1.21 |
| 4 | 3,6 | 1.35 | 1.40 |
| 3 | 4,6 | 1.29 | 1.42 |
| Palmityl methyl ester | | 34.40 | |

**Table 6.** Retention time of $O$-acetyl-$O$-methyl-derivatives of methyl D-galactosides and D-mannosides relative
to palmityl methyl ester.

| Position of substituent | | Retention times (min) | | |
|---|---|---|---|---|
| | | Galactose | | Mannose |
| Methyl | Acetyl | α | β | α |
| 2,3,4,6 | | 0.39 | 0.36 | 0.35 |
| 2,3,4 | 6 | 0.562 | 0.561 | 0.53 |
| 2,3,6 | 4 | 0.48 | 0.47 | 0.54 |
| 2,4,6 | 3 | 0.59 | 0.54 | 0.51 |
| 3,4,6 | 2 | 0.535 | 0.65 | 0.44 |
| 2,3 | 4,6 | 0.68 | 0.79 | 0.78 |
| 2,4 | 3,6 | 0.80 | 0.76 | 0.74 |
| 2,6 | 3,4 | 0.73 | 0.70 | 0.68 |
| 3,4 | 2,6 | 0.77 | 0.856 | 0.64 |
| 3,6 | 2,4 | 0.64 | 0.71 | 0.63 |
| 4,6 | 2,3 | 0.705 | 0.73 | 0.65 |
| 2 | 3,4,6 | 0.89 | 0.86 | 0.89 |
| 3 | 2,4,6 | 0.85 | 0.91 | 0.85 |
| 4 | 2,3,6 | 0.91 | 0.96 | 0.93 |
| 6 | 2,3,4 | 0.80 | 0.82 | 0.80 |
| Palmityl methyl ester | | | 34.30 | |

**Table 7.** Main fragments leading to the characterization of partially methylated and acetylated methylglycosides.

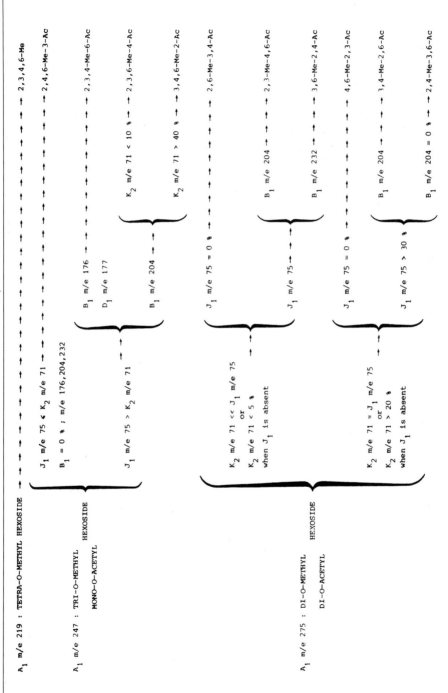

**Table 7.** continued.

$A_1$ m/e 303 : **MONO-O-METHYL HEXOSIDE TRI-O-ACETYL**

$K_2$ m/e 71 > 20 %
- $J_1$ m/e 75 > 60 % → 3-Me-2,4,6-Ac
- $J_1$ m/e 75 < 5 %
  - $E_1$ m/e 289 → 6-Me-2,3,4-Ac
  - $H_1$ m/e 116 ≈ 0 %
  - $H_1$ m/e 116 > 15 % → 2-Me-3,4,6-Ac

$K_2$ m/e 71 < 5 % → 4-Me-2,3,6-Ac

$F_1$ m/e 142 : 4-O-METHYL GLUCOSAMINIDE

$B_1$ m/e 217
- $A_2$ m/e 228 → 3,4,6-Me

$B_1$ m/e 217 (or 245) absent
- $A_2$ m/e 256 → 3,4-Me-6-Ac
- m/e 45 > 30 %
- $H_1$ m/e 116 = 0 % → 4,6-Me-3-Ac
- m/e 45 < 5 %
- $H_1$ m/e 116 > 10 % → 4-Me-3,6-Ac

$F_1$ m/e 170 : 4-O-ACETYL GLYCOSAMINIDE

$B_1$ m/e 245
- $A_2$ m/e 228 → 3,6-Me-4-Ac

$B_1$ π/e 245 absent
- $A_1$ m/e 316 → 3-Me-4,6-Ac
- $A_2$ m/e 284 → 6-Me-3,4-Ac

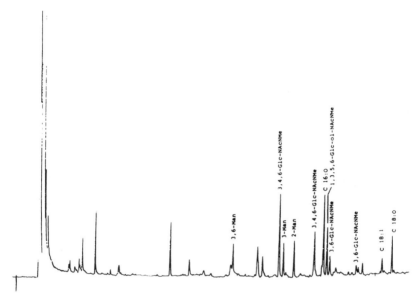

**Figure 17.** G.l.c. (glass capillary column wall coated with OV101; flame ionization detection) of partially methylated and aceylated methylglycosides obtained by methanolysis of 10 μg of permethylated oligosaccharide-alditol 11 from hen ovomucoid (see *Figure 9*). 1.5 μg of the sample was injected.

derivatives of methyl-2-deoxy-2-(*N*-methyl acetamido)-D-glucopyranoside, methyl-D-galactoside and D-mannoside are given in *Tables 5* and *6*, respectively (see also Chapter 1, Section 7).

E.I. mass spectra are analysed according to Zolotarev *et al.* (50).

*Table 7* gives the main fragments leading to the characterization of partially methylated and acetylated methylglycosides (51).

*Figure 17* gives the g.l.c. pattern of partially methylated and acetylated methylglycosides obtained after methylation, with lithium methyl sulfinyl carbanion reagent, and methanolysis of 10 μg of oligosaccharide alditol 11 from hen ovomucoid. It can be observed that g.l.c. on a Silicone OV 101 capillary column with flame ionization detector allows identification of all methyl ethers without use of m.s. so showing that the lithium methylsulphinyl carbanion reagent gives less background in g.l.c. Analysis of samples containing 1 μg or less of total sugars necessitates the use of chemical ionization (ionizing gas: ammonia) with specific ions: $(M + 18)^+$ for neutral methyl ethers (m/e 268 for permethyl derivatives; m/e 296 for tri-*O*-methyl-mono-*O*-acetyl derivatives; m/e 324 for di-*O*-methyl-di-*O*-acetyl derivatives; m/e 352 for mono-*O*-methyl-tri-*O*-acetyl derivatives); $(M + 1)^+$ for hexosamine methyl ethers (m/e 292 for permethyl derivatives; m/e 320 for di-*O*-methyl-mono-*O*-acetyl derivatives; m/e 348 for mono-*O*-methyl-di-*O*-acetyl derivatives); $(M + 1)^+$ for methyl ethers of *N*-acetylneuraminic acid (m/e 408).

## 7.4 Sequence and molecular weight determination of permethylated oligosaccharides by e.i.-m.s. and f.a.b.-m.s.

Direct analysis of permethylated oligosaccharides by electron impact mass spectrometry (e.i.-m.s.) and fast atom bombardment mass spectrometry (f.a.b.-m.s.) provides in-

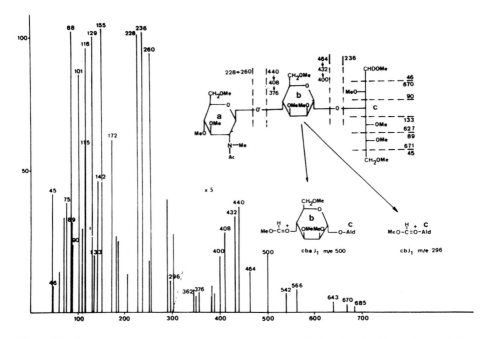

**Figure 18.** Mass spectrum of permethylated GlcNAc($\beta$1-4)Man($\alpha$1-3)Man-ol obtained by partial acetolysis of glycopeptide GP V-5 from $\alpha_1$-acid glycoprotein.

teresting information on the monosaccharide sequences within oligosaccharide chains as well as on their molecular weight.

### 7.4.1 *E.i.-m.s. of permethylated oligosaccharides*

(i) *Gas liquid chromatography:* Permethylated disaccharides to pentasaccharides can be analysed by g.l.c. coupled to a mass spectrometer. G.l.c. is performed under the following conditions in our laboratory: Girdel gas chromatograph (Model 300); glass columns (100 cm × 0.3 cm i.d.) packed with Supelcoport (100 – 120 mesh) containing 1% Dexsil 300 (column temperature: 150 – 320°C with a temperature gradient of 4°C/min; flow rate of carrier gas (helium): 40 ml/min), or capillary column of OV-101 (60 m × 0.4 mm i.d.) (column temperature: 140 – 320°C with a temperature gradient of 5°C/min; flow rate of carrier gas (helium): 5 ml/min).

(ii) *E.i.-mass spectrometry:* We use a Riber-Mag 10-10 mass spectrometer (Rueil-Malmaison, France) under the following experimental conditions: electron energy: 70 eV, ionization current: 0.2 mA; ionization chamber temperature: 190°C. *Figure 18* gives a typical mass spectrum of a permethylated trisaccharide-alditol obtained by partial acetolysis of glycopeptide GP V-5 from $\alpha_1$-acid glycoprotein. The ion observed at m/e 260 is formed by the cleavage of a hexosamine residue from the oligosaccharide, while subsequent elimination of methanol gives rise to m/e 228. The fragment at m/e 236 originates by splitting off the alditol residue from this oligosaccharide. The ions observed at m/e 464, 432, 400 and 440, 408, 376 establish the sequence of the monosaccharides in the trisaccharide as being GlcNAc-Man-Man. In addition, the ion at m/e

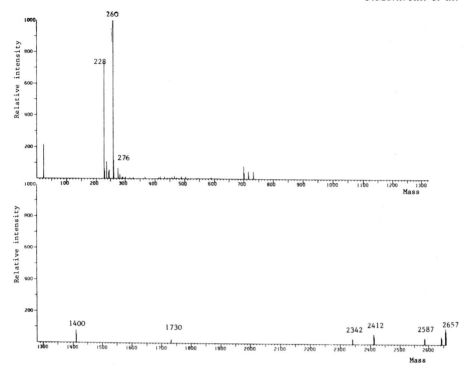

**Figure 19.** F.a.b. spectrum of oligosaccharide-alditol 11 obtained by hydrazinolysis from hen ovomucoid (see *Figure 9*).

133 demonstrates that the alditol unit is glycosylated in C-3 position, and the ion at m/e 296 proves that the linkage of hexosamine to the hexose unit is not a 1,3-linkage. Since the sum of the intensities of the ions at m/e 440, 408 and 376 is larger than that of the ions observed at m/e 464, 432 and 400, the type of linkage is a 1,4- or 1,6-linkage rather than an 1,2-linkage. Moreover, the relative abundance of the ions at m/e 440, 408 and 376, decreases from high to low m/e values suggesting an 1,4-linkage of the hexosamine to the hexose residue.

### 7.4.2 *f.a.b.-m.s. of permethylated oligosaccharides*

F.a.b.-m.s. of permethylated sugar derivatives in acheived by direct injection of the samples into the mass spectrometer (VG analytical ZAB-HF, reversed geometry).

(i)     Dissolve the samples in methanol to a concentration of about 5 $\mu g/\mu l$.

(ii)    Load the stainless steel target with 1 $\mu l$ of a 0.1% (w/v) solution of sodium acetate in methanol.

(iii)   Dry and apply to the target $3-4$ $\mu l$ of 1-mercapto-2,3-propanediol (EGA Chemie, Steinheim, FRG) and 0.5 to 1.0 $\mu l$ of sample solution.

(iv)    Bombard the target with xenon atoms having a kinetic energy equivalent to $9.0-9.5$ kV.

(v)     Record the spectra routinely at 7 kV acceleration voltage giving a mass range of about 3800 amu in a mass-controlled scan. Record the spectra in the positive ion mode by a mass controlled scan of $100-500$ sec duration depending on the Mr of the sample.

*Figure 19* gives the f.a.b. spectrum of an oligosaccharide-alditol obtained by hydrazinolysis (see Section 2.3) of hen ovomucoid. It shows intense pseudomolecular ions $M+Na^+$ at m/e 2657 and ions at m/e 276 and 260 produced by the alditol residue and by the non-reducing terminal monosaccharide, respectively. The ion at m/e 2342 is produced by the cleavage of the glycosidic bond of the chitobiose unit with the positive charge remaining on the large non-reducing oligosaccharide fragment.

F.a.b.-m.s. allows the detection of accompanying homologues that are not easily revealed by other spectroscopic or degradative methods. For example, the ion at m/e 2412 indicates the presence of a lower homologue formed by the loss of one N-acetyl-D-glucosamine residue from the native oligosaccharide. The ion at m/e 2342 is produced by fission of the chitobiose linkage. In this component, it can be concluded that one of GlcNAc residue of the chitobiose moiety has been certainly lost during the hydrazinolysis process (see Section 2.3.3).

## 8. USE OF GLYCOSIDASES

### 8.1 Experimental approach

Glycosidases are excellent tools for (a) the elucidation of the primary structure of glycans by sequential degradation; (b) the determination of the anomeric linkage of each conjugated monosaccharide; (c) the controlled modification of glycoprotein glycans, including membrane glycoproteins, in order to explore their biological role; and (d) the preparation of specific acceptors for glycosyltransferase activity studies. Basically, two types of enzymes are used: *exoglycosidases* which hydrolyse glycosidic bonds of monosaccharides in terminal non-reducing positions and may achieve a stepwise degradation of the glycan; and *endoglycosidases* which have been recently introduced and rapidly developed because of their very promising performances. They hydrolyse internal glycosidic bonds, so liberating oligosaccharides or the glycans themselves. In *Table 8* are reported the origin and the commercial sources of glycosidases.

The use of glycosidases was limited for a long time by the following factors:

(i)    Glycosidase activities depend on the origin of the enzyme (microorganisms, plants or animals, cytoplasm, lysosome or Golgi apparatus), on the type of glycoside linkage ($1\rightarrow2$, $1\rightarrow3$, $1\rightarrow4$ or $1\rightarrow6$), on the nature of the substrate (artificial substrates or natural substrates like oligosaccharides, glycopeptides or glycoproteins), on the structure of substrates (branched, phosphorylated or sulphated) and on the spatial conformation (i.e. mono- to penta-antennary glycans).

(ii)   Commercial sources of monospecific and pure glycosidases are relatively rare. In many cases, it is necessary to verify the absence of contaminant enzymes like proteases, phospholipases or other glycosidases, and often to purify the commercial materials. In other cases, the lack of active enzymes has led the experimentors to prepare their own glycosidases.

(iii)  At the present time, only neuraminidases are available commercially in an immobilised form although immobilized glycosidases would be of interest in the production of hydrolysed substrates free from contaminating glycosidases and thus, avoiding artifacts due to their glycoprotein nature.

**Table 8.** Origin of glycosidases.

| Sources[a] | Exoglycosidases[b] | | | | | | | Endoglycosidases[b] | | |
|---|---|---|---|---|---|---|---|---|---|---|
| | 1 | 2 | 3 | 4 | 5 | 6 | 7 | 8 | 9 | 10 |
| *Vibrio cholerae* (a-d,i) | + | | | | | | | | | |
| *Arthrobacter ureafasciens* (d,h) | + | | | | | | | | | |
| *Arthrobacter sialophilus* | + | | | | | | | | | |
| *Clostridium perfringens* (e-g) | + | | | | | | | $+(C_I,C_{II})$ | | |
| *Streptococcus pneumoniae* (e,j,k) | + | + | | + | | | | $+(D)$ | | + |
| *Streptococcus sanguis* | + | | | | | | | | | |
| *Bacillus fulminans* | | | | | | | + | | | |
| *Bacteroides fragilis* | + | | | | | | | | | + |
| *Escherichia coli* (d-g,i-k) | | + | + | | | | | | | + |
| *Escherichia freundii* (j) | | | | | | | | | | + |
| *Flavobacterium meningosepticum* | | | | | | | | $+(F)$ | + | |
| Baker yeast | | | | | + | | | | | |
| *Kluiveromyces fragilis* (e) | | + | | | | | | | | |
| *Streptomyces plicatus* (e,j) | | | | | | | | $+(H)$ | | |
| *Aspergillus niger* (e) | | + | + | + | + | + | + | | | |
| *Aspergillus oryzae*(e) | | + | + | + | | | | | | |
| *Polyporus sulfureus* | | | | | | + | | | | |
| *Sporotrichum dimorphosporum* | | | | | | | | $+(B)$ | | |
| *Mortierella vinacea* | | | + | | | | | | | |
| *Trichomonas foetus* | + | | | | | | | | | |
| *Canavalia ensiformis* (e) | | + | | + | + | | | | | |
| Coffee bean | | | + | | | | | | | |
| *Phaseolus vulgaris* | | | | | + | | | | | |
| Pineapple stem bromelain | | | | | + | + | | | | |
| Almond emulsin (e) | | + | | | + | | + | | + | |
| Fig | | | | | | | | $+(F_I,F_{II})$ | | |
| *Turbo cornutus* (j,k) | | + | + | + | + | + | + | | | |
| *Charonia lampas* (e,j,k) | | | | | | | + | | | |
| Snail (k) | | | | | + | | | | | |
| *Helix pomatia* | | | | | | + | + | | | |
| Hen oviduct | | | | | + | + | + | + | | |
| Mammalian tissues: liver (e) | | + | + | + | | | + | + | | |
| kidney (d,e) | | | | + | + | | + | + | | |
| epididymis (e) | | | | + | | | | | | |
| placenta (e) | | | | + | | | | | | |

[a]Suppliers: a: General Biochemicals (Chagrin Falls, OH, USA); b: Behringwerke (Marburg/Lahn, FRG); c: Calbiochem-Behring Corporation (La Jolla, CA, USA); d: Boehringer (Mannheim, FRG); e: Sigma Chemical Company (St Louis, MO, USA); f: Worthington Biochemical Corporation, a division of Millipore Corporation (Freehold, NJ, USA); g: P.L. Biochemical, GmbH (St Goar, FRG); h: Nakarai Chemical Ltd (Kyoto, Japan); i: Koch Light Ltd (Haverhill, Suffolk, UK); j: Miles (Naperville, IL, USA); k: Seikagaku Kogyo Co (Tokyo, Japan).
[b]Glycosidases: 1: Neuraminidase; 2: $\beta$-D-galactosidase; 3: $\alpha$-D-galactosidase; 4: $N$-acetyl-$\beta$-D-hexosaminidase; 5: $\alpha$-D-mannosidase; 6: $\beta$-D-mannosidase; 7: $\alpha$-L-fucosidase; 8: endo-$N$-acetyl-$\beta$-D-glucosaminidase; 9: $N$-glycopeptide hydrolase; 10: endo-$\beta$-D-galactosidase.

(iv)  Glycosidase activities are generally detected and measured by using synthetic *p*-nitrophenyl or methylumbelliferyl-sugar derivatives as substrates. This could be a cause of error since some glycosidases which are active on synthetic substrates are not active on natural structures, and *vice-versa*. Before listing the properties of differents glycosidases used in the study of glycoproteins, we shall give some technical terms and experimental procedures used in enzymic methods.

### 8.1.1 *General definitions*

*Enzyme unit (U)*: amount of enzyme which liberates 1 $\mu$mol of monosaccharide (exoglycosidase) or oligosaccharide (endoglycosidase) in 1 min under well defined conditions of pH, temperature and effector(s).

*Specific activity*: defined in U/mg of protein.

### 8.1.3 *General methods*

*Use of nitrophenyl derivatives as substrates*

*Reagents*

(A)    10 mM nitrophenyl-glycoside in aqueous solution.
(B)    McIlvaine buffer (0.2 M sodium phosphate-0.1 M citric acid; pH adjusted to the optimal pH).
(C)    1 M sodium carbonate aqueous solution.

*Method*

(i)     Add 0.1 ml of nitrophenylglycoside to 0.1 ml of McIlvaine buffer and 0.2 ml enzyme solution.
(ii)    Incubate for 5 — 15 min at the optimal temperature (previously determined) and stop the reaction by adding 0.6 ml of the sodium carbonate solution.
(iii)   Determine the optical density of the liberated nitrophenol at 400 nm.

*Use of methylumbelliferyl derivatives as substrates*

*Reagents*

(A)    1 mM methylumbelliferyl glycoside in aqueous solution.
(B)    McIlvaine buffer (see above).
(C)    0.2 M glycine buffer adjusted to pH 10.7 with 1 M NaOH.

*Method*

Apply the same procedure as above and determine the amount of liberated methylumbelliferon by spectrofluorimetry (excitation $\lambda$: 390 nm; emission: 450 nm).

## 8.2 **Exoglycosidases**

### 8.2.1 *Neuraminidase (EC: 3.2.1.18; N-acetylneuraminyl hydrolase)*

*Origin of enzyme*: In spite of the existence of numerous sources of enzyme (*Table 8*), only neuraminidases from *Vibrio cholerae*, *Clostridium perfringens* and *Arthrobacter ureafasciens* are commercially available. Their properties are described in *Table 9*.

*Method*

(i)     Dissolve the substrate in 50 — 100 $\mu$l of 50 mM sodium acetate buffer at the optimum pH of enzyme (in the case of the neuraminidase from *Vibrio cholerae*

**Table 9.** Properties of neuraminidases.

| Origin of enzyme | Optimum pH | $Km^a$ (mM) | Effector | pI | Mw (kDa) |
|---|---|---|---|---|---|
| *Vibrio cholerae* | 5.6 | 1 | $Ca^{2+}$ | 4.8 | 66 |
| *Clostridium perfringens* | $4.3 - 5.2$ | 0.2 | No | 5.1 | 63 |
| *Arthrobacter ureafasciens* | $5 - 5.5^b$ | 0.6 | No | – | $39 - 51$ |

[a]Substrate: $\alpha$-2,3-*N*-acetylneuraminyllactose (Neu5Ac$\alpha$-2,3 lactose).
[b]$4.3 - 4.5$ with colominic acid as substrate.

the buffer contains 10 mM $CaCl_2$ and 0.1 M NaCl in addition).

(ii)    Add 0.1 U of neuraminidase per $\mu$mol of *N*-acetylneuraminic acid to be removed.

(iii)    Incubate for from 10 min to 48 h at 37°C, the time of hydrolysis depending on (a) the nature of the substrate (oligosaccharide, glycopeptide or glycoprotein), (b) the type of linkage ($\alpha$-2,3, $\alpha$-2,6 or $\alpha$-2,8) and (c) the spatial conformation of glycans or glycoproteins (mucin type or globular protein type).

(iv)    Stop the hydrolysis by heating in a boiling water bath for 3 min.

(v)    Determine the amount of liberated sialic acid by applying colorimetric methods (see Section 5.4.2 and Chapter 1, Section 2.7).

*General comments*

(a)    Enzyme specificity (for reviews, see Ref. 52 and 53). Neuraminidases from bacterial origin are generally able to split $\alpha$-2,3 and $\alpha$-2,6 linkages easily. Enzymes from viruses (FPV, NDV or A2) act only on $\alpha$-2,3 linkages. In the case of *Arthrobacter ureafasciens*, the enzyme splits $\alpha$-2,6 linkages much faster than $\alpha$-2,3 linkages. The $\alpha$-2,8 linkages of colominic acid form a poor substrate for bacterial neuraminidases. The rate of hydrolysis depends on the substitution of neuraminic acid. *N*-glycolylneuraminyl linkages are resistant to hydrolysis except in the case of the enzyme from *Clostridium perfringens* which is able to split both $\alpha$-2,3-*N*-acetyl and *N*-glycolylneuraminyl linkages. *O*-substitution generally causes a 50% decrease of the enzyme activity, except in the case of enzyme from *Streptococcus sanguis* which splits 7-, 8- or 9-*O*-acetyl-*N*-acetylneuraminic acid at the same rate as sialyllactose, but removes the 4-*O*-acetyl derivative 5-fold slower, similar to the *Vibrio cholerae* enzyme.

(b)    Sulphate groups decrease the rate of hydrolysis. However, sulphation of the monosaccharide adjacent to that with which *N*-acetylneuraminic acid is linked, increases the rate of hydrolysis.

(c)    *N*-acetylneuraminyl bonds are cleaved differently according to the monosaccharide on which the *N*-acetylneuraminic acid is linked. For example, sialyl *N*-acetylgalactosamine linkages are relatively resistant to neuraminidase hydrolysis.

(d)    The specificity of commercial neuraminidases is given in *Table 10*.

### 8.2.2 $\beta$-D-Galactosidase (EC: 3.2.1.23; $\beta$-D-galactoside galactohydrolase)

*Origin of enzyme*. The most popular $\beta$-D-galactosidase preparations are from *Canavalia ensiformis* (Jack bean) (54) and almond emulsin (55). Other sources are given in *Table 8*.

**Table 10.** Specificity of neuraminidase (results expressed as % of total hydrolysis).

| Origin of enzyme | Neu5Ac | | | Neu5Gc | Bovine submaxillary mucin | Human serum transferrin | α₁-Acid glycoprotein |
|---|---|---|---|---|---|---|---|
| | α-2,3 | α-2,6 | α-2,8 | α-2,3 | | | |
| Vibrio cholerae | 100 | 53 | 31 | 25 | 55 | 100 | 100 |
| Clostridium perfringens | 100 | 44 | 44 | 20 | 20 | 100 | 100 |
| Arthrobacter ureafasciens | 63 | 100 | 44 | 8 | 52 | 100 | ND |

ND: not determined.

*Method using Canavalia ensiformis enzyme.*

(i)     Dissolve the substrate in $50-100$ $\mu$l of 0.02 M sodium phosphate-0.01 M acetic acid buffer pH 3.5 (the pH can be increased to pH 4.0 when the substrate is not stable at the more acidic pH).

(ii)    Add 0.4 U of $\beta$-D-galactosidase per $\mu$mol of galactose to be liberated. The final volume should not exceed 400 $\mu$l.

(iii)   Incubate the mixture at 37°C from 10 min to 48 h.

(iv)    Stop the hydrolysis by heating in a boiling water-bath for 3 min.

(v)     Determine the free galactose by conventional methods such as h.p.l.c., g.l.c. or the galactose dehydrogenase assay (see Chapter 1, Section 2.8).

*Comments*

(a)     The enzyme activity is stable for 5 days at 37°C.

(b)     Glycopeptides and glycoproteins are hydrolysed 20-fold and 150-fold slower than oligosaccharides, respectively.

(c)     The enzyme can act at high ionic strengths but not in chaotropic medium.

(d)     Two groups of enzymes can be characterized. The first group present a high affinity toward the lactose (e.g. the enzymes from *E. coli, K. lactis* or *K. fragilis*), are activated by $Na^+$, $K^+$, $Mg^{2+}$ or $Co^{2+}$ ions and do not act on glycoconjugate derivatives. The second group are active on the $\beta$-galactosidic linkages present in glycoconjugates.

(e)     Generally, the $\beta$-D-galactosidases do not act when the monosaccharide to which the galactose is conjugated, is substituted as in lacto-*N*-fucopentaose II:

$$\text{Gal}(\beta1\text{-}3)[\text{Fuc}(\alpha1\text{-}4)]\ \text{GlcNAc}(\beta1\text{-}3)\text{Gal}(\beta1\text{-}4)\text{Glc}$$

(f)     Some $\beta$-D-galactosidases present a very strict specificity. For example, the enzyme from *S. pneumoniae* hydrolyses $\text{Gal}(\beta1\text{-}4)\text{GlcNAc}$ and lactose but not $\text{Gal}(\beta1\text{-}3)\text{GlcNAc}$ or $\text{Gal}(\beta1\text{-}6)\text{GlcNAc}$.

(g)     When the concentration of substrate is low, the enzymes from almond emulsin and Jack bean hydrolyse the $\beta$-1,4 linkages 13-fold and 50-fold faster than $\beta$-1,3 linkages, respectively. On the contrary, when the substrate concentration is high, there is no difference between the rate of hydrolysis by both enzymes. In the same way, bovine epididymis enzyme cleaves $\beta$-1,3 linkages more easily than $\beta$-1,4 or $\beta$-1,6 ones while the reverse situation is observed for calf intestine $\beta$-galactosidase. In *Table 11* the activities of $\beta$-galactosidase from Jack bean and *E. coli* toward different substrates are compared.

**Table 11.** Specificity of $\beta$-D-galactosidases from jack bean and *E. coli* (results expressed as % of total hydrolysis).

| Substrates | Jack bean | E. coli |
|---|---|---|
| Gal($\beta$1-6)GlcNAc | 100 | 100 |
| Gal($\beta$1-4)GlcNAc | 75 | 55 |
| Gal($\beta$1-3)GlcNAc | 1 | 92 |
| Gal($\beta$1-4)Glc | 42 | 100 |
| Gal($\beta$1-6)Man | 19 | 0 |
| Asialofetuin | 100 | – |
| $\alpha_1$-Acid glycoprotein | 79 | 0 |

**Table 12.** Specificity of *N*-acetyl-$\beta$-D-hexosaminidases (results expressed as % of total hydrolysis).

| Substrates | Jack bean | Aspergillus niger |
|---|---|---|
| GlcNAc($\beta$1-2)Man | 100 | 62 |
| GlcNAc($\beta$1-4)Man | 35 | 56 |
| GlcNAc($\beta$1-6)Man | 23 | 100 |
| GlcNAc($\beta$1-4)Gal | 23 | 28 |
| GlcNAc($\beta$1-3)Gal($\beta$1-4)Glc | 68 | 40 |
| GlcNAc($\beta$1-2)Man($\alpha$1-3)Man | 47 | 53 |
| Man($\alpha$1-3)[GlcNAc($\beta$1-4)]Man | 81 | 0 |
| GlcNAc($\beta$1-4)Man($\alpha$1-3)[GlcNAc($\beta$1-4)]Man | 71 | 0 |
| Asialo-agalacto serotransferrin glycopeptide | 80 | – |

### 8.2.3. $\alpha$-D-Galactosidase (EC: 3.2.1.22; $\alpha$-D-galactoside galactohydrolase)

Use of $\alpha$-D-galactosidases is limited to glycoproteins having blood-group activity and to other rare glycoproteins posessing $\alpha$-galactosyl residues. Experimental procedure is identical as for $\beta$-D-galactosidase. The principal sources of enzyme are listed in *Table 8*.

### 8.2.4 N-acetyl-$\beta$-D-hexosaminidase (EC: 3.2.1.52; 2-acetamido-2-deoxy-$\beta$-D-hexoside acetamidodeoxyhexohydrolase)

*Origin of enzyme.* Generally, *N*-acetyl-$\beta$-D-hexosaminidases act on *N*-acetyl-$\beta$-D-glucosaminyl and *N*-acetyl-$\beta$-D-galactosaminyl linkages, the ratio of activities depending upon the origin of enzyme. The sources of *N*-acetyl-hexosaminidases are listed in *Table 8*, the enzyme from Jack bean being the most utilized (56).

*Method.*
(i)     Dissolve 1 $\mu$mol of substrate in $100-200$ $\mu$l of McIlvaine buffer pH 5.0.
(ii)    Add $1.5-5$ U of enzyme in $50-100$ $\mu$l of buffer.
(iii)   Incubate for 6 to 48 h at 37°C.
(iv)    Stop the reaction by heating in a boiling water-bath for 3 min.

The enzyme is inhibited by acetate buffer. Enzyme activity is destroyed by denaturing medium. *N*-acetyl-$\beta$-D-hexosaminidase activity is stable 3 days at 37°C. The rate of hydrolysis depends on the origin of enzyme, the type of linkage and on the molecular conformation of the molecule (*Table 12*). The enzyme is more active on oligosaccharides than on glycopeptides and glycoproteins.

### 8.2.5 $\alpha$-D-Mannosidase (EC: 3.2.1.24; $\alpha$-D-mannoside mannohydrolase)

*Origin of enzyme: Table 8* summarizes the most important sources of $\alpha$-D-mannosidases. The most utilized is the enzyme from Jack bean (51).

*Method*
(i)     Dissolve the substrate in 50 mM sodium acetate buffer pH 4.2.
(ii)    Add the enzyme in the ratio of 1.5 U/$\mu$mol of mannose to be liberated.
(iii)   Incubate the mixture from 30 min to 48 h at 37°C.

(iv)     Stop the hydrolysis by heating in a boiling water-bath for 3 min.

All $\alpha$-D-mannosidases, except *Aspergillus niger* enzyme, require from 0.1 to 5 mM of $Zn^{2+}$ salts in order to stabilize the enzyme toward pH and temperature. Enzyme activity is maximal in the presence of at least 0.01% (w/v) total proteins in the incubating medium. The Jack bean enzyme hydrolyses Man($\alpha$1-2)Man or Man($\alpha$1-6)Man linkages easily, but the Man($\alpha$1-3)Man linkage is cleaved 15-fold slower. The *Aspergillus niger* enzyme is very active on Man($\alpha$1-6)Man, Man($\alpha$1-4)Man or Man($\alpha$1-4)GlcNAc. It is inactive on Man($\alpha$1-3)Man but weakly active on Man($\alpha$1-2)Man. Golgian $\alpha$-mannosidases participating in the maturation of *N*-glycosylproteins glycans and having a very sharp specificity ($\alpha$-1,2-mannosidases; $\alpha$-1,3 and $\alpha$-1-6-mannosidases, GlcNAc-dependent) have been described.

### 8.2.6 $\beta$-D-Mannosidase (EC: 3.2.1.25; $\beta$-D-mannoside mannohydrolase)

*Origin of enzyme.* No preparations are commercially available and *Table 8* summarizes the most important sources of enzymes described in the literature. We currently use a $\beta$-D-mannosidase we isolated from *A. niger* and which is, at the moment, the most active one described (53).

### Method

(i)     Dissolve 1 $\mu$mol of substrate in 100 $\mu$l of 20 mM sodium-phosphate buffer (pH 3.5), add 5 mU of *A. niger* $\beta$-D-mannosidase (ensure that the final volume does not exceed 200 $\mu$l).

(ii)     Incubate the mixture at 37°C for $1-24$ h.

(iii)     Stop the reaction by heating in a boiling water-bath for 3 min.

Some general comments on this procedure are:

(a)     The enzyme is specific for $\beta$-1,4-mannosyl linkages (*Table 13*).

(b)     Glycans and glycopeptides are hydrolysed 4- to 5-fold faster than synthetic substrates.

(c)     The rate of hydrolysis decreases when a fucose residue is $\alpha$-1,6-linked to the *N*-acetylglucosamine residue of *N*-glycosylamine linkage.

### 8.2.7 $\alpha$-L-Fucosidase (EC: 3.2.1.51; $\alpha$-L-fucoside fucohydrolase)

*Origin of enzyme.* The enzyme is widely distributed in various living organisms (*Table*

**Table 13.** Specificity of $\beta$-D-mannosidases (results expressed as % of total hydrolysis).

| Substrates | Hen oviduct | Tremella fusiformis | A. niger |
|---|---|---|---|
| *p*-Nitrophenyl-$\beta$-D-mannoside | 100 | 100 | 100 |
| Man($\beta$1-4)GlcNAc | 33 | 10 | 100 |
| Man($\beta$1-3)GlcNAc | – | – | 0.04 |
| Man($\beta$1-6)GlcNAc | – | – | 0.1 |
| Man($\beta$1-4)GlcNAc($\beta$1-4)GlcNAc($\beta$1-N)Asn | 25 | – | 100 |
| Man($\beta$1-4)GlcNAc($\beta$1-4)[Fuc($\alpha$1-6)]GlcNAc($\beta$1-N)Asn | – | – | 80 |

8) and can be isolated in a pure form by affinity chromatography on fucosylamine-Sepharose columns.

*Method*

(i)     Dissolve the substrate in 50 $\mu$l of 100 mM phosphate-citrate buffer at the enzyme's optimum pH.

(ii)     Add $50-100$ $\mu$l of the enzyme preparation in the ratio of 1 U of $\alpha$-L-fucosidase/$\mu$mol of fucose to be liberated.

(iii)     Incubate the mixture at 37°C from 3 to 48 h.

(iv)     Stop the reaction by heating in a boiling water-bath for 3 min.

Some general comments on this procedure are:

(a)     Fucosidases having a broad activity spectrum are weakly active toward fucose residues $\alpha$-1,6-conjugated to the asparagine-linked *N*-acetylglucosamine. They are unable to remove this kind of fucose residue when acting on native glycoproteins (for fucosidase specificity, see ref. 57 and 60).

(b)     Free fucose is a powerful competitor inhibitor (Ki = 100 $\mu$M).

(c)     The rate of hydrolysis depends on the origin of the enzyme. For example, increasing activities of *Turbo cornutus* enzyme are as follows: Fuc($\alpha$1-4)GlcNAc > Fuc($\alpha$1-2)Gal > Fuc($\alpha$1-3)GlcNAc, and Fuc($\alpha$1-2)Gal > Fuc($\alpha$1-4)GlcNAc > Fuc($\alpha$1-3)GlcNAc for *Charonia lampas* enzyme.

(d)     In some cases, $\alpha$-L-fucosidase is highly specific for $\alpha$-1,2-fucosyl linkage (e.g. *Clostridium perfringens, Aspergillus niger* or *Streptococcus sanguis*).

(e)     Fucosidases from microorganisms are able to remove fucose from native glycoproteins but are inactive toward synthetic substrates. On the contrary, mammalian enzymes act on synthetic substrates, oligosaccharides and glycopeptides but do not act on native glycoproteins.

(f)     The specificity of fucosidases of different origins is described in *Table 14*.

**Table 14.** Specificity of $\alpha$-L-fucosidases (results expressed as % of total hydrolysis).

| Substrates | Clostridium perfringens | Almond | Mammalian tissues |
|---|---|---|---|
| *p*-Nitrophenyl $\alpha$-L-fucoside | 0 | – | 100 |
| Fuc($\alpha$1-2)Gal | 100 | – | – |
| Fuc($\alpha$1-2)Fuc | 0 | – | – |
| Fuc($\alpha$1-2)Gal($\beta$1-4)Glc | 79 | 1.5 | 28 |
| Fuc($\alpha$1-3)Gal($\beta$1-4)Glc | – | 13 | – |
| Fuc($\alpha$1-2)Gal($\beta$1-4)[Fuc($\alpha$1-3)]Glc | 39 | – | – |
| Lacto-*N*-fucopentaose I | 80 | 1 | 2 |
| Lacto-*N*-difucohexaose I | 7 | 17 | – |
| Lacto-*N*-fucopentaose II | 4 | 100 | 20 |
| Lacto-*N*-fucopentaose III | 9 | 93 | 6 |
| IgG glycopeptide | 0 | 0 | 54 |
| Porcine submaxillary mucin | 90 | – | – |
| Canine submaxillary mucin | 90 | – | – |
| Human ovarian cyst glycoprotein | 0 | – | – |
| H-antigen glycolipid | – | – | 100 |

### 8.3 **Endoglycosidases**

Endoglycosidases which release oligosaccharides from conjugated glycans are divided into two classes. In *class I*, the enzymes split the monosaccharide-amino acid or monosaccharide-peptide linkage, removing the complete glycan (aspartyl-*N*-acetylglucosaminidase; peptide-*N*-glycosidase). In *class II*, the enzymes cleaves sugar-sugar glycosidic bonds, liberating part of the glycan (endo-*N*-acetyl-*β*-D-glucosaminidase or endo-*β*-D-galactosidase).

### 8.3.1 *4-N-(2-β-D-glucosaminyl)-L-asparaginase (EC: 3.5.1.26; 4-N-(2-acetamido 2-deoxy-β-D-glucopyranosyl-L-asparagine aminohydrolase)*

*Origin of enzyme.* The enzyme is of lysosomal origin and hydrolyses glycoasparagines liberating the glycan, aspartic acid and ammonia (61).

*Method*

(i)     Dissolve $3-20$ μmol of glycoasparagine in 50 μl of 0.1 M phosphate buffer at the optimum pH.

(ii)    Add $0.5-5$ mU of enzyme and incubate the mixture (final volume 100 μl) at 37°C for from 30 min to 24 h.

(iii)   Stop the hydrolysis by heating in a boiling water-bath for 3 min.

    The enzyme hydrolyses only the *N*-acetylgucosaminyl−asparagine linkage of glyco-asparagines and is inactive on glycopeptides. It is active on glycoasparagines of the oligomannosidic and of the *N*-acetyllactosaminic type. The structure and size of the carbohydrate moieties does not seem to affect the enzyme activity.

### 8.3.2 *Peptide-N-glycosidase*

Peptide-*N*-glycosidase hydrolyzes the GlcNAc-Asn bond not only of glycoasparagines, in a similar manner to the above enzyme, but also of glycopeptides and some glyco-proteins.

*Origin of enzyme.* The enzyme is present in almond emulsin (62) and in the culture medium of *Flavobacterium meningosepticum* (63) (see *Table 8*).

*Method*

(A) Use of peptide *N*-glycosidase from almond emulsin.

(i)     Add $5-20$ U of enzyme to 1 μmol of glycopeptide or 100 μg of glycoproteins (native or denatured) dissolved in 0.2 M phosphate-citrate buffer, pH 5.1.

(ii)    Incubate the mixture at 37°C for from 30 min to 24 h in the presence of pepstatin (1 μg), phenyl-methylsulfonyl fluoride (10 mM) and iodacetic acid (10 mM).

(iii)   Stop the reaction by heating the solution in a boiling water-bath for 3 min.

(B) Use of peptide-*N*-glycosidase from *F. meningosepticum*.

(i)     Incubate 1 μmol of glycopeptide for from $2-24$ h at 37°C in the presence of 30 U of enzyme in 0.1 M glycine-sodium hydroxide pH 9.3.

(ii)    Stop the hydrolysis by addition of 1 M HCl (1 vol/3 vol of the mixture).

    Some general comments on this procedure are:

(a) Three types of peptide-*N*-glycosidases have been characterised. Types A and B are active on large glycopeptides (up to 11 amino acid residues) while type C is active on glycopeptides containing less than 4 amino-acid residues. Some glycoproteins are hydrolysed by type A only if desialylated.

(b) *N*-acetyllactosaminic type glycopeptides are hydrolyzed faster than hybrid or oligomannosidic type glycopeptides.

(c) The presence of fucose does not decrease the rate of hydrolysis.

(d) The enzyme is stable in 1 M NaSCN chaotropic medium for 48 h at 37°C.

### 8.3.3 *Endo-β-galactosidases*

Endo-β-D-galactosidases cleaves internal β-galactoside linkages as follows:

$$Gal(\beta1\text{-}4)GlcNAc(\beta1\text{-}3)Gal(\beta1\text{-}4)(monosaccharide)_n + H_2O$$
$$\rightarrow Gal(\beta1\text{-}4)GlcNAc(\beta1\text{-}3)Gal + (monosaccharide)_n$$

Origin of enzyme: The enzyme has been found only in the microorganisms *Streptococcus pneumoniae*, *Escherichia freundii* and *Bacteroides fragilis* (see *Table 8*).

### Method

(i) Dissolve 5 − 10 nmol of substrate in 100 μl of 50 mM sodium acetate buffer pH 5.8.

(ii) Add 5 − 20 mU of enzyme (64) and incubate at 37°C for from 1 to 24 h. When using the enzyme from *Bacteroides fragilis*, stabilize activity with 0.02% (w/v) serum albumin.

(iii) Stop the reaction by heating in a boiling water-bath for 3 min.

The enzymes are unable to split the galactosyl linkages when galactose residues are sulphated. Fucosylated glycans are less hydrolysed. The substitution by fucose at the C-2 position of the terminal galactose residues makes the substrate resistant to the enzyme. Similarly, the enzyme is inactive when galactose residues are substituted at the position C-6 by *N*-acetylglucosamine to form a branched-point. Substitution of the terminal galactose residue at the C-3 position by *N*-acetylneuraminic acid or by α-galactosyl residues increases the enzyme activity.

### 8.3.4 *Endo-N-acetyl-β-D-glucosaminidases*

Endo-*N*-acetyl-β-D-glucosaminidases constitute an homogeneous family of enzymes from various sources (see *Table 8*) that act on the β-1,4-*N*-acetylglucosaminyl bond of the *N*,*N'*-diacetylchitobiose residue present in all *N*-glycosylproteins. The definition of their specificity, described in *Table 15* allows them to be classified in three groups.

*Group I* comprises of the so-called 'endo-H (65), '-C$_{II}$' (66) and '-F$_{II}$' which are specific of oligomannosidic and hybrid type glycans.

*Group II* is represented by the 'endo-D' (65) and '-C$_{II}$' (66) which require an unsubstituted α-1,3-mannose in terminal non-reducing position or substituted only in C-4 position. These enzymes act poorly on glycopeptides of the oligomannosidic type.

*Group III* comprises the 'endo-B' (68) and '-F' (69) which are active on glycans of both the *N*-acetyllactosaminic and oligomannosidic types, but not on glycans of the hybrid type.

**Table 15.** Specificity of endo-*N*-acetyl-β-D-glucosaminidases.

| Substrates | Endo-*N*-acetyl-β-D-glucosaminidase | | | |
|---|---|---|---|---|
| | H | D | $C_{II}$ | B |
| *N*-acetyllactosaminic type structures[a] | | | | |
| NeuAc(α2-6)Gal(β1-4)GlcNAc(β1-2)Man(α1-3)— NeuAc(α2-6)Gal(β1-4)GlcNAc(β1-2)Man(α1-6)— >Man(β1-4)R₁ | − | − | − | − |
| NeuAc(α2-6)Gal(β1-4)GlcNAc(β1-2)Man(α1-3)— Gal(β1-4)GlcNAc(β1-2)Man(α1-6)— >Man(β1-4)R₁ | − | − | − | + |
| Gal(β1-4)GlcNAc(β1-2)Man(α1-3)— Gal(β1-4)GlcNAc(β1-2)Man(α1-6)— >Man(β1-4)R₁ | − | − | − | + |
| Gal(β1-4)GlcNAc(β1-1)Man(α1-3)— Gal(β1-4)GlcNAc(β1-2)Man(α1-6)— >Man(β1-4)R₂ | − | − | − | + |
| GlcNAc(β1-2)Man(α1-3)— Man(α1-6)— >Man(β1-4)R₂ | − | − | − | + |
| GlcNAc(β1-4)Man(α1-3)— Man(α1-6)— >Man(β1-4)R₂ | − | + | − | + |
| Asialotriantennary glycopeptide[b] | − | − | − | + |
| Asialotetraantennary glycopeptide[c] | − | − | − | − |
| Oligomannosidic type structures | | | | |
| Man(α1-3)— Man(α1-6)— >Man(β1-4)R₁ | + | + | − | + |
| Man(α1-3)— Man(α1-3)Man(α1-6)— >Man(β1-4)R₁ | + | − | + | + |
| Man(α1-3)\ Man(α1-3)—Man(α1-6)— >Man(β1-4)R₁ Man(α1-6)/ | + | + | + | + |
| Man(α1-2)Man(α1-3)\ Man(α1-3)———Man(α1-6)— >Man(β1-4)R₁ Man(α1-6)— | + | − | + | + |
| Man(α1-2)Man(α1-2)Man(α1-3)— [Man(α1-2)]Man(α1-3)\Man(α1-6)/ >Man(β1-4)R₁ Man(α1-2)Man(α1-6)/ | + | − | + | + |
| Hybrid type structure | + | − | − | − |

[a]R₁: GlcNAc(β1-4)GlcNAc(β-N)Asn or peptide; R₂: GlcNAc(β1-4)[Fuc(α1-6)]GlcNAc(β-N)Asn or peptide.
[b,c]: desialylated structures 5 and 7, respectively, from *Figure 5*.
[d]: As an example, see *Figure 7*.

### 8.3.5 Endo-*N*-acetyl-β-D-glucosaminidase H (EC: 3.2.1.96; mannosyl glycoprotein 1,4-*N*-acetamido-deoxy-β-D-glycohydrolase)

*Origin of enzyme.* The enzyme from the culture medium of *Streptomyces plicatus* is the most currently used of the enzymes listed in *Table 8*.

### Method

(i)    Dissolve the substrate in 50 mM sodium citrate buffer pH 5.5.

(ii)    Add endoglycosidase H so as to obtain a substrate/endoglycosidase ratio of 50:1 (w/w).

(iii)   Incubate the mixture for 24 h.

(iv)    Stop the reaction by heating in a boiling water-bath for 3 min.

Some general comments on this procedure are:

(a)     The enzyme is not inhibited by the presence of $\alpha$-1,6-linked fucose in the core of hybrid structures.

(b)     The activity does not decrease when phosphates, glucose or *N*-acetylglucosamine residues are present.

(c)     The activity of proteases which are always present in endo-H-preparations can be reduced by the addition of PMSF at a final concentration of 1 to 2 mM.

(d)     Endo-H activity is stable in 0.2% (w/v) SDS medium and retains 30% of its activity in 1% (w/v) SDS medium.

(e)     The activity of the enzyme is stable in 0.5 M NaSCN for 24 h. In addition, NaSCN inhibits the proteolytic activity associated to endo-H preparations.

(f)     Details concerning the optimization of hydrolysis by endo-H from *Streptomyces plicatus* are given in ref. 66.

### 8.3.6 *Endo-N-acetyl-β-D-glucosaminidase (EC: 3.2.1.x; glycoprotein 1,4-N-acetamido-deoxy-β-D-glycohydrolase)*

This enzyme, discovered by our laboratory in the culture medium of *Sporotrichum dimorphosporum*, cleaves the di-*N*-acetylchitobiose residue of *N*-glycosidically linked glycans of both oligomannosidic and *N*-acetyllactosaminic types.

*Origin of enzyme*. The preparation of enzyme from culture medium of *Sporotrichum dimorphosporum* is described in detail in ref. 68.

*Method:*

(i)     Dissolve 20 nmol of substrate in 50 $\mu$l of 0.02 M phosphate-0.01 M citrate buffer pH 5.0 and add 5 mU of enzyme.

(ii)    Incubate the mixture at 50°C for from 10 min to 3 h in the case of glycopeptides and at 37°C for from 6 h to 24 h for glycoproteins.

(iii)   Stop the reaction by heating in a boiling water-bath for 3 min or by ethanol precipitation (final concentration of ethanol: 50% v/v).

Some general comments on this procedure are:

(a)     Substitution by a fucose residue $\alpha$-1,6-linked to the terminal *N*-acetylglucosamine residue decreases the rate of hydrolysis.

(b)     The enzyme is more active toward glycopeptides than glycoproteins and oligosaccharides.

(c)     In order to ensure complete deglycosylation it is generally important to denaturate the glycoprotein by heating prior to enzymic hydrolysis.

## 9. ACKNOWLEDGEMENTS

We would like to express our sincere gratitude to Prof. André Verbert for his critical comments. We wish to thank Mrs Gisèle Tinel and Brigitte Layous-Mahieu for excellent technical assistance in the typing of the manuscript.

## 10. REFERENCES

1. Montreuil,J. (1980) *Adv. Carbohydr. Chem. Biochem.*, **37**, 157.
2. Montreuil,J. (1982) in *Comprehensive Biochemistry*. Neuberger,A. and Van Deenen,L.L.M. (eds.). Elsevier, Amsterdam, Vol. 19B, Part II, p. 1.
3. Montreuil,J. (1984) *Pure Appl. Chem.*, **56**, 859.
4. Montreuil,J. (1984) *Biol. Cell.*, **51**, 115.
5. Kobata,.A (1984) in *Biology of Carbohydrates*, Ginsburg,V. and Robbins,P.W. (eds.). John Wiley, New York, Vol. 2, p. 87.
6. Hughes,C. (1983) *Glycoproteins*, Chapman and Hall, London.
7. Vliegenthart,J.F.G., Van Halbeck,H. and Dorland,L. (1981) *Pure Appl. Chem.*, **53**, 45.
8. Vliegenthart,J.F.G., Dorland,L. and Van Halbeck,H. (1983) *Adv. Carbohydr. Chem. Biochem.*, **41**, 209.
9. Dell,A., Morris,H.R., Egge,H., Von Nicolai,H. and Strecker,G. (1983) *Carbohydr. Res.*, **115**, 41.
10. Kamerling,J.P., Heerma,W., Vliegenthart,J.F.G., Green,B.N., Lewis,I.A.S., Strecker,G. and Spik,G. (1983) *Biomed. Mass Spectr.*, **10**, 420.
11. Yamashita,K., Mizuochi,T. and Kobata,A. (1982) in *Methods in Enzymology*, Ginsburg,V. (ed.). Academic Press Inc., New York, Vol. 83, p. 105.
12. Iyer,R.N. and Carlson,D.M. (1971) *Arch. Biochem. Biophys.*, **142**, 101.
13. Lee,Y.C. and Scocca,J.R. (1972) *J. Biol. Chem.*, **247**, 5753.
14. Yosizawa,Z., Sato,T. and Schmid,K. (1966) *Biophys. Biochim. Acta*, **121**, 417.
15. Dische,Z. (1955) in *Methods Biochem. Anal.* Glick,D. (ed.), Interscience, New York, Vol. 2, p. 313.
16. Montreuil,J. and Spik,G. (1963) *Méthodes colorimétriques de dosage des glucides totaux*, Monog. Lab. Chim. Biol. Fac. Sci. Lille.
17. White,C.A. and Kennedy,J.F. (1981) in *Techniques in Carbohydrate Metabolism*, Kornberg,H.L., Metcalfe,J.C., Northcote,D.H., Pogson,C.I. and Tipton,K.F. (eds.) Elsevier, Amsterdam, Vol. B3, B312.
18. Montreuil,J., Spik,G., Fournet,B. and Tollier,M.T. (1981) in *Techniques d'Analyse et de Contrôle dans les Industries Agro-Alimentaires*, Deymié,B., Multon,J.L. and Simon,D. (ed.) Technique et Documentation, Paris, Vol. 4, p. 85.
19. Schauer,R. (1978) in *Methods in Enzymology*, Ginsburg,V. (ed.), Academic Press, New York, Vol. 50, p. 64.
20. Green,J. (1981) in *Techniques in Carbohydrate Metabolism*, Kornberg,H.L., Metcalfe,J.C., Northcote,D.H., Pogson,C.I. and Tipton,K.F. (eds.), Elsevier, Amsterdam, Vol. B3, B311.
21. Baenziger,J.U. and Natowicz,M. (1981) *Anal. Biochem.*, **112**, 357.
22. Bergh,M.L.E., Koppen,P. and Van den Eijnden,D.H.(1981) *Carbohydr. Res.*, **94**, 225.
23. Blumberg,K., Liniere,F., Pustilnik,L. and Bush,C.A. (1982) *Anal. Biochem.*, **119**, 409.
24. Lis,H. and Sharon,N. (1977) in *The Antigens*, Sela,M. (ed.), Academic Press, New York, Vol. 4, p. 429.
25. Goldstein,I.J. and Hayes,C.E. (1978) *Adv. Carbohydr. Chem. Biochem.*, **35**, 127.
26. Blake,D.A. and Goldstein,I.J. (1983) in *Methods In Enzymology*, Ginsburg,V. (ed.), Academic Press, New York, Vol. 83, p. 127.
27. Lis,H. and Sharon,N. (1984) in *Biology of Carbohydrates*, Ginsburg,V. and Robbins,P. (eds.), J. Wiley, New York, p. 1.
28. Montreuil,J., Debray,H., Debeire,P. and Delannoy,P. (1983) in *Structural Carbohydrates in the Liver*, Popper,H., Reutter,W., Gudat,F. and Köttgen,E. (eds.), MTP Press, London, Falk Symposium, **34**, p. 239.
29. Kornfeld,R. and Ferres,C. (1975) *J. Biol. Chem.*, **250**, 2614.
30. Debray,H., Decout,D., Strecker,G., Spik,G. and Montreuil,J. (1981) *Eur. J. Biochem.*, **117**, 41.
31. Debray,H., Pierce-Crétel,A., Spik,G. and Montreuil,J. (1983) in *Lectins – Biology, Biochemistry, Clinical Biochemistry*, Bøg-Hansen,T.C. and Spengler,G.A. (eds.), De Gruyter, Berlin, Vol. 3, p. 335.
32. March,S.C., Parikh,I. and Cuatrecasas,P. (1974) *Anal. Biochem.*, **60**, 149.
33. Wilchek,M. and Miron,T. (1974) *Mol. Cell. Biochem.*, **4**, 181.
34. Dulaney,J.T. (1979) *Mol. Cell. Biochem.*, **21**, 43.
35. Lotan,R. and Nicolson,G.L. (1979) *Biochim. Biophys. Acta*, **559**, 329.
36. Bøg-Hansen,T.C. (1973) *Anal. Biochem.*, **56**, 480.
37. Lotan,R., Gussin,A.E.S., Lis,H. and Sharon,N. (1973) *Biochem. Biophys. Res. Commun.*, **52**, 656.
38. Toyoshima,S., Osawa,T. and Tonomura,A. (1970) *Biochim. Biophys. Acta*, **221**, 514.
39. Murphy,L.A. and Goldstein,I.J. (1977) *J. Biol. Chem.*, **252**, 4739.

40. Eckhardt,A.E. and Goldstein,I.J. (1983) *Biochemistry*, **22**, 5290.
41. Shankar Yyer,P.N., Wilkinson,K.D. and Goldstein,I.J. (1976) *Arch. Biochem. Biophys.*, **177**, 330.
42. Yamashita,K., Kochibe,N., Ohkura,T., Veda,I. and Kobata,A. (1985) *J. Biol. Chem.*, **260**, 4688.
43. Cummings,R.D. and Kornfeld,S. (1982) *J. Biol. Chem.*, **257**, 11235.
44. Gombas,G. and Zanetta,J.-P. (1978) in *Research Methods in Neurochemistry*, Marks,N. and Rodnight,R. (eds.), Vol. 4, p. 307.
45. Montreuil,J., Spik,G., Fournet,B. and Tollier,M.T. (1981) in *Techniques d'Analyse et de Contrôle dans les Industries Agro-Alimentaires*, Deymié,B., Multon,J.L. and Simon,D. (ed.) Technique et Documentation, Paris, Vol. 4, p. 102.
46. Lindberg,B. (1972) in *Methods in Enzymology*, Ginsburg,V. (ed.), Academic Press, New York, Vol. 28, p. 173.
47. Lindberg,B. and Lönngren,J. (1978) in *Methods in Enzymology*, Ginsburg,V. (ed.), Academic Press, New York, Vol. 50, p. 3.
48. Hakomori,S.I. (1964) *J. Biochem.* (Tokyo), **55**, 205.
49. Paz Parente,J., Cardon,P., Leroy,Y., Montreuil,J., Fournet,B. and Ricart,G. (1985) *Carbohydr. Res.*, **141**, 41.
50. Zolotaren,B.M., Ott,A.Ya and Chizhov,O.S. (1978) *Adv. Mass Spectrom.*, **7B**, 1371.
51. Fournet,B., Strecker,G., Leroy,Y. and Montreuil,J. (1981) *Anal. Biochem.*, **116**, 489.
52. Corfield,A.P., Michalski,J.-C. and Schauer,R. (1981) *Perspect. Inherited Metab. Dis.*, **4**, 3.
53. Schauer,R. (1982) *Adv. Carbohydr. Chem. Biochem.*, **40**, 131.
54. Li,S.C., Mazotia,M.Y., Chien,S.F. and Li,Y.T. (1975) *J. Biol. Chem.*, **250**, 6786.
55. Arakawa,M., Ogata,S.I., Muramatsu,T. and Kobata,A. (1974) *J. Biochem.* (Tokyo), **75**, 707.
56. Li,S.C. and Li,Y.T. (1970) *J. Biol. Chem.*, **245**, 5153.
57. Tai,T., Yamashita,K., Ogata,A., Koide,N., Muramatsu,T., Iwashita,S., Inoue,Y. and Kobata,A. (1975) *J. Biol. Chem.*, **250**, 8569.
58. Bouquelet,S., Spik,G. and Montreuil,J. (1978) *Biochim. Biophys. Acta*, **522**, 521.
59. Nishigaki,M., Muramatsu,T., Kobata,A. and Maeyama,K.I. (1974) *J. Biochem.* (Tokyo), **75**, 509.
60. Ogata-Arakawa,M., Muramatsu,T. and Kobata,A. (1977) *Arch. Biochem. Biophys.*, **181**, 353.
61. Dugal,B. (1978) *Biochem. J.*, **171**, 799.
62. Tarentino,A.L. and Plummer,T.H.,Jr. (1982) *J. Biol. Chem.*, **257**, 10776.
63. Plummer,T.H.,Jr., Elder,J.H., Alexander,S., Whelan,A.W. and Tarentino,A.L. (1984) *J. Biol. Chem.*, **259**, 10700.
64. Scudder,P., Hanfland,P., Uemura,K.I. and Feizi,T. (1984) *J. Biol. Chem.*, **259**, 6586.
65. Trimble,R.B. and Maley,F. (1984) *Anal. Biochem.*, **141**, 515.
66. Ito,J., Muramatsu,T. and Kobata,A. (1975) *Arch. Biochem. Biophys.*, **171**, 78.
67. Koide,N. and Muramatsu,T. (1974) *J. Biol. Chem.*, **249**, 4897.
68. Bouquelet,S., Strecker,G., Montreuil,J. and Spik,G. (1980) *Biochimie*, **62**, 43.
69. Elder,J.H. and Alexander,S. (1982) *Proc. Natl. Acad. Sci. USA*, **79**, 4540.
70. Blumenkrantz,N. and Asboe-Hansen,G. (1973) *Anal. Biochem.*, **54**, 484.
71. Belcher,R., Nutten,A.J. and Sambrook,C.M. (1954) *Analyst*, **79**, 201.
72. Niazi,S. and State,D. (1948) *Cancer Res.*, **8**, 653.

CHAPTER 6

# Glycolipids

IAN M.MORRISON

## 1. INTRODUCTION

Glycolipid is the general term for a wide range of compounds, usually of biological significance, which are distributed throughout animal, plant and microbial cells. The carbohydrate, or glyco-, components are oligosaccharide chains which are frequently branched and can also carry non-carbohydrate substituents such as acetyl or sulphate groups. It is these precise carbohydrate structures which confer the biological specificity to a particular glycolipid.

The lipid portion can be far more variable and some authorities suggest that any hydrophobic compound which can combine glycosidically with a carbohydrate can be considered in this context. More commonly, the classification of lipids is reserved for derivatives of glycerol, sterols, isoprenols and sphingenine-type bases. The glycerol and sphingenine-type bases are further substituted by long-chain saturated and unsaturated fatty acids. The glycerol moiety is substituted at C-1 and C-2 by fatty acids linked by ester bonds (C-3 of the glycerol unit is always reserved for the carbohydrate chain) and different fatty acids can be present while the sphingenine-type bases are substituted at their amino-function by the fatty acids. The same oligosaccharide chain and sphingenine base can be substituted by different fatty acids in the one species. *N*-acyl sphingenine is called ceramide while the term cerebroside refers to a monoglycosyl-ceramide.

The isoprenol type of glycolipids are not present in great amounts but are very widespread in nature acting as transporters for the carbohydrate components of biologically important molecules.

Glycerol-based glycolipids are also well distributed and are by far the major type of glycolipid in the plant kingdom. The ceramide family is found mainly in animal and microbial tissues. As well as neutral glycosphingolipids, there are important families of sulphoglycosphingolipids and glycosphingolipids which contain sialic acid residues (commonly called gangliosides). These animal glycolipids are now known to be vital for blood group typing and antigenicity as well as their role in many metabolic disorders.

The analysis of glycolipids presents a number of problems. Care must be taken to ensure that all substituent groups, especially those more labile under acidic or alkaline conditions, are retained by the molecule and do not become detached or migrate during the extraction and purification procedures. In addition, since the carbohydrate component has hydrophilic properties due to the large number of hydroxy groups present and the lipid portion is generally hydrophobic, the ambivalent state of these compounds causes extraction and separation difficulties due to solubility effects. For more detailed structural investigations of glycolipids, emphasis will be placed on the carbohydrate component. The detailed chemistry of the lipid components will not be considered in any rigorous fashion; only where there is a gross change in the nature of the lipid component (e.g. glycoglycerolipids and glycoceramides with the same carbohydrate chains) will attention be paid to the finer details and physiological and biochemical implications.

## 2. SEPARATION METHODS

### 2.1 Extraction of glycolipids

Most emphasis in this and other sections will be on the animal glycolipids since these are studied in far greater detail at present. The extraction etc. of plant glycolipids can be carried out by very similar methods to those outlined below: indeed the extraction and purification of plant glycolipids is less complex since the oligosaccharide chains are usually shorter than in many of the animal glycolipids making them more soluble in lipophilic solvents.

The extraction method which has gained most acceptance over the last 30 years is that introduced by Folch *et al.* (1). If at all possible fresh tissue should be used and not dried tissue. The extractant is chloroform:methanol (2:1 v/v) and at least 20 volumes is used per gram of tissue, extraction being performed in a Waring Blender for 3 min in the cold. If only dried tissue is available, longer extraction periods of up to 24 h at room temperature with continuous stirring are required. The insoluble residue, comprising mainly proteinaceous and other non-lipid components, is removed either by filtration through a layer of celite or sintered glass or by centrifugation. Filtration through paper is avoided as sialic acid-containing glycolipids are adsorbed. Further quantities of glycolipid can be obtained from the residue from the Folch procedure by homogenization with 10 volumes of chloroform:methanol (1:2 v/v) to which is added 5% water. This second extraction, proposed by Suzuki (2) is essential when polyglycosyl glycolipids and more polar gangliosides are present. A modification of this second extraction is used by Slomiany and Slomiany (3) where hog gastric mucosa (50 g), after the Folch extraction, is further extracted by stirring with 2000 ml of 0.4 M sodium acetate in chloroform:methanol:water (30:15:4 v/v/v) for 24 h at room temperature. After filtering, the residue was extracted with a further 1000 ml of the same solvent. This solvent extracts the bulk of the gangliosides.

Chloroform:methanol mixtures can be replaced by phosphate buffer/tetrahydrofuran (1:8 v/v) in a procedure reported by Tettamanti *et al.* (4). The tissue (1 g) is homogenized in a high-speed blender with 1 ml of 0.01 M potassium phosphate buffer (pH 6.8) for 1 min. Tetrahydrofuran (8 ml) is added and homogenization continued for a further 1 min. After centrifugation at 12 000 *g* for 10 min, at 15°C, the pellet is re-extracted with 1 ml of the phosphate buffer and 4 ml of tetrahydrofuran for 2 min and re-centrifuged. The extraction is repeated twice and the gangliosides are located in the supernatant solution.

### 2.2 Purification of glycolipids

#### 2.2.1 *Partition into upper and lower phases*

No single method can be used to separate glycolipids: the method employed depends on the glycolipids under consideration. The glycolipids in the above extracts are contaminated with non-lipids which have some characteristics similar to the glycolipids, such as glycoproteins, and also by other lipid classes.

Starting with the chloroform-methanol extracts obtained from either the Folch and/or Suzuki methods, add 0.2 volumes of 0.12 M NaCl or KCl and the extract partitions into two phases. The lower phase contains the neutral glycolipids with relatively short carbohydrate chains along with contaminating neutral lipids and phospholipids. The upper phase contains most of the gangliosides and glycolipids with longer carbohydrate chains. Glycoproteins or glycopeptides may also be present in this phase. The separation is not absolute as the phase in which a particular glycolipid appears will depend on its structure. On separation of the phases, each phase should be backwashed with the solvent of the other phase and the phase and washings of the same composition combined.

In the procedure involving aqueous tetrahydrofuran, which was specifically designed for the extraction of gangliosides, 0.3 volumes of diethyl ether is added to the extract and after centrifuging at 600 $g$ for 20 min, at 15°C, the gangliosides are recovered from the aqueous phase.

### 2.2.2 *Other methods of purification*

The aqueous phase from the Folch fractionation procedure can be concentrated on a rotary evaporator and dialysed at $2-4$°C for 3 days against distilled water. The impure ganglioside mixture, retained in the bag, is lyophilized and stored at $-20$°C.

The total extract can be chromatographed on a 100 × 2 cm column of Sephadex LH20. Using an eluant of chloroform:methanol:water (40:20:1 v/v/v), the total lipid is eluted at the void volume of the column and non-lipids are retained (5).

Non-lipids are also removed by chromatography on small columns (15 × 1.0 cm) of Sephadex G25 in the bead form (6). The preparation of this column is more complex. A chloroform:methanol:water (8:4:3 v/v/v) mixture is prepared which gives an upper and a lower phase. The gel is swollen and packed in the upper phase then the lower phase is used for final washing and the chromatographic separation. The glycolipids are again eluted in the void volume of this system.

## 2.3 **Separation of glycolipid mixtures**

### 2.3.1 *Chromatography on DEAE-matrices*

Convenient methods of separating acidic glycolipids (gangliosides) from neutral glycolipids is via DEAE-cellulose, DEAE-Sephadex or even DEAE-silica gel.

DEAE-cellulose, dry and in the hydroxide form, is suspended in glacial acetic acid and packed, batchwise, in the same solvent into a 30 × 3 cm column (7). Such a column is suitable for the separation of approximately 500 mg of the glycolipid mixture. The packed column is washed with methanol and chloroform:methanol (2:1 v/v) to remove the acetic acid and finally with chloroform:methanol:water (450:50:1 v/v/v) before applying the sample in the last solvent. The neutral glycolipids are partially purified by stepwise elution with solvents containing progressively lower proportions of chloroform until methanol alone is used. The gangliosides can then be eluted with chloroform:methanol:0.05 M ammonium acetate (4:1:1 v/v/v) followed finally by glacial acetic acid.

With improved t.l.c. and h.p.l.c. methods for identifying individual glycolipids in mixtures, a simplified separation of gangliosides from neutral glycolipids on DEAE-Sephadex A-25 has many advantages (8). The matrix (2.2 g) is generated in the acetate form by washing with 0.8 M sodium acetate in methanol and packed in a 1.4 cm column. The column is washed with chloroform:methanol:water (15:30:4 v/v/v) and a sample (from 100 ml plasma) applied in the same solvent. Washing with this solvent elutes the neutral glycolipids while subsequent washing with 0.8 M sodium acetate in methanol removes the gangliosides.

In all procedures with DEAE-matrices, salts used to elute the gangliosides are removed by evaporation of the organic solvents and dialysis.

Since gangliosides can contain a number of sialic acid residues, gradient elution of DEAE-Sephadex columns will produce mono-, di- and trisialo-ganglioside fractions (9). A crude ganglioside mixture (2−5 g) is dispersed in chloroform:methanol (1:1 v/v, 60 ml) and mixed with DEAE-Sephadex A-25 (acetate form: 5 g) with vigorous shaking and then left for 20 min. The suspension is diluted with methanol (30 ml) and water (8 ml) and packed in a 27 × 4.6 cm column. The column is first washed with three column volumes of chloroform:methanol:water (15:30:4 v/v/v) to remove neutral lipids and then with one column volume of methanol to remove neutral glycolipids. The gangliosides were fractionated by applying a concave gradient from 0 to 0.45 M ammonium acetate in methanol. This gradient was achieved by connecting, in series, three vessels containing 1200 ml respectively of 0.05 M, 0.15 M and 0.45 M ammonium acetate in methanol. The higher the molarity of the ammonium acetate, the greater will be the number of sialic acid residues in the molecule.

### 2.3.2 *Chromatography on silica gel*

Glycolipid mixtures, partially purified by previous methods, are further fractionated on silica gel columns from various manufacturers. No specific type of silica gel is necessarily the best: different types are recommended for different separations. Elution of the column is with chloroform:methanol mixtures and gradient elution should be used to give better separations.

Many mixtures of glycolipids are too polar to be separated without tailing on the silica gel columns. These glycolipids are acetylated by the addition of pyridine:acetic anhydride (2:1 v/v, 1.0 ml) at room temperature and left overnight (10). The solvents are removed on a rotary evaporator at a bath temperature of 40°C and final traces are removed by repeated addition of toluene and further evaporation. The acetylated glycolipids are then separated on the silica gel column and eluted with chloroform:methanol mixtures. The individual glycolipids are deacetylated with 0.1% sodium methoxide dissolved in 1.0 ml chloroform:methanol (2:1 v/v). Although there are few reports of problems with this method, care should be taken to ensure that the acetylation and deacetylation steps do not cause any hydrolysis of chemical bonds or migration of unstable substituent groups.

### 2.3.3 *Thin layer chromatography*

Silica gel plates are recommended for separation of all types of glycolipids. In general, neutral solvent systems such as chloroform:methanol:water (65:25:4 v/v/v) give better results with neutral glycolipids. Basic systems such as chloroform:methanol:ammonia solution (sp. grav. 0.88):water (60:35:1:7 v/v/v/v) or those containing $Ca^{2+}$ ions, such as chloroform:methanol:0.02% aqueous calcium chloride (60:40:9 v/v/v) give better separation for gangliosides.

Contrary to what might be expected, acetylation of glycolipids does not give better separation on t.l.c. Indeed, mixtures of glycolipids with the same amino sugar content will migrate at the same rate.

When carrying out preparative t.l.c., plates should not be allowed to dry out as it is almost impossible to elute the appropriate bands with suitable solvents even with sonication.

Bands or spots on t.l.c. plates can be detected by conventional spray reagents used for non-reducing sugars such as diphenylamine-aniline-$H_3PO_4$ or naphthoresorcinol-$H_2SO_4$ (see Chapter 1, *Table 3*): these reagents will be able to detect both native and acetylated glycolipids.

High-performance t.l.c. on superfine (5 $\mu$m) silica gel has been used to separate complex mixtures of gangliosides and neutral glycolipids (11).

### 2.3.4 *High-performance liquid chromatography*

Some important separations have been achieved by h.p.l.c., however the detection methods available are not very sensitive for native glycolipids and large amounts are required. Most of the better separations have involved derivatized glycolipids, the derivatives used allowing detection in the u.v. region. They are usually benzoyl or *p*-nitrobenzoyl esters and the columns are based on silica gel. As an example, the glycolipid ($\sim 0.1 - 1.0$ mg) is dissolved in 20% benzoyl chloride in 0.6 ml of dry pyridine in a screw-capped vial fitted with a Teflon-coated liner and heated at 60°C for 1 h. The solvent is evaporated off in a stream of $N_2$. The residue is dissolved in hexane (5 ml) and washed successively with 95% methanol saturated with sodium carbonate, 95% methanol, 0.6 M hydrochloric acid in 95% methanol and 95% methanol. The hexane layer is finally evaporated to dryness in a stream of $N_2$ and taken up in a small volume of hexane for injection into the liquid chromatograph. The stainless steel column (80 cm × 2.1 mm i.d.) was packed with the silica gel, Zipax (E.I.du Pont and Nemours and Co. Inc.) and the liquid phase was a linear gradient system of 0.20% methanol to 0.75% methanol in hexane. The detector was operated at 254 nm and the minimum detection level was about 10 pmol of each glycolipid. By this method, the benzoyl derivatives of mono-, di-, tri- and tetrahexosyl ceramides were separated in 25 min (12).

The *p*-nitrobenzoyl derivatives are prepared in the same way. The *p*-nitrobenzoyl derivatives are more sensitive to u.v. detection than the benzoyl derivatives but the benzoyl derivatives are separated more clearly than the *p*-nitrobenzoyl derivatives in all the systems tried.

A potential drawback to this method is the problem of regenerating the native glycolipid. It is not too difficult to remove *O*-benzoyl groups but the *N*-benzoyl group on amino sugars and in the ceramide fragment are far more difficult to remove. However, the use of a catalyst such as 4-dimethylaminopyridine in the benzoylation reaction with benzoic anhydride produces only the *O*-benzoyl substitution. In this procedure, the glycolipid (0.1 − 1.0 mg) is treated with a freshly prepared solution of 20% benzoic anhydride and 5% 4-dimethylaminopyridine in 0.5 ml dry pyridine and left at 37°C for 4 h. The products are recovered as above except that 80% methanol is used instead of 95% methanol. The *O*-benzoyl substituents are removed by treating with 1.0 ml of 0.5 M sodium hydroxide in methanol at 37°C for 1 h. Chloroform:methanol:water (8:4:3 v/v/v, 5 ml) is added and the lower phase contains the regenerated glycolipid (13).

## 3. CHEMICAL METHODS OF ANALYSIS

### 3.1 Composition of glycolipids

#### 3.1.1 *Carbohydrate components*

(i) Hydrolyse 50−200 μg of glycolipid by heating at 80°C for 6 h in 0.3 ml of glacial acetic acid: 0.15 M sulphuric acid (9:1 v/v) in a Teflon lined screw-capped vial.

(ii) Cool the hydrolysate and pass through a 200 mg Dowex 1-X8 (acetate) resin column washed with methanol.

(iii) Evaporate the filtrate to dryness under $N_2$.

(iv) Reduce the sugars with 10 mg of sodium borohydride in 0.3 ml of water.

(v) After 3 h add two drops of acetic acid and evaporate the solution to dryness. Repeat the evaporation five times with 5% acetic acid in methanol to remove borate ion as the volatile methyl borate.

(vi) Acetylate the alditols by heating with 0.3 ml of acetic anhydride at 100°C for 2 h and remove the excess acetic acid by repeated evaporations with toluene.

(vii) Partition the residue between chloroform and water (1:1 v/v, 5 ml of each), remove the chloroform layer and dry (14).

The acetylated alditols are analysed by g.l.c. using methods outlined in detail in Chapter 1, Section 6.2.2, which also described an alternative derivatization procedure.

Methanolysis presents an alternative method of cleavage.

(i) Dissolve 50−200 μg of glycolipid in 1.0 ml of 1.0 M methanolic hydrogen chloride (prepared by bubbling 3.65 g hydrogen chloride gas into 100 ml dry methanol) and heat at 85°C for 18 h. Methanolysis is carried out in sealed tubes under $N_2$.

(ii) Add 50 mg silver carbonate to neutralise the acid and 50 μl acetic anhydride to re-*N*-acetylate amino sugars.

(iii) Leave the mixture for 6 h and centrifuge.

(iv) Remove the methanolic solution and wash the residue three times with methanol (1.0 ml).

(v) Evaporate the methanolic solution to dryness.

(vi) Trimethylsilylate the methyl glycosides by the addition of 0.05 ml of a mixture of dry pyridine:hexamethyldisilazane:trimethylchlorosilane (5:1:1 v/v/v) and leave at room temperature for 30 min.

(vii) Analyse the TMS-methyl glycosides by g.l.c. using procedures outlined in Chapter 1, Section 6.2.1 (15), which also described an alternative methanolysis procedure.

It is usually necessary to use both types of hydrolysis since the conditions of methanolysis do not fully release the amino sugars from the glycolipids while the acidic hydrolysis conditions cause partial destruction of the sialic acid residues.

### 3.1.2 *Fatty acid components.*

(i)     Hydrolyse 2−5 mg of glycolipid by heating with 2.0 ml of a mixture of conc. HCl:methanol:$H_2O$ (3:29:4 v/v/v) at 78°C for 18 h in a sealed tube under $N_2$ (16).

(ii)    Extract the released fatty acids and their methyl esters with 3 × 2.0 ml of petroleum ether (BP 40−60°C) and evaporate the extracts to dryness.

(iii)   Treat the extract with 1.0 ml at 1.5 M methanolic hydrogen chloride (prepared by bubbling 5.5 g hydrogen chloride gas into 100 ml dry methanol) to esterify the fatty acids.

(iv)    Heat for 2 h at 40°C in a sealed tube, add 1 ml of water and extract the esters into hexane (3 × 2 ml).

(v)     Wash the hexane solution with 2% potassium carbonate and dry over anhydrous sodium sulphate.

If hydroxy fatty acids are found to be present, trimethylsilylate the hydroxy group as described for the carbohydrate analysis to give faster separation on g.l.c. with less tailing of peaks.

G.l.c. analysis can be carried out on a variety of columns but the following method is recommended. The packing used is 15% EGSS-Y on Chromosorb W (100−120 mesh:AW:DMCS treated). The glass columns are 2 m × 4 mm i.d. with carrier gas of $N_2$ (50 ml min$^{-1}$) and operated at 194°C. EGSS-X can be used at 178°C to give faster analysis but the separations are not always as good.

### 3.1.3 *Sphingosine base components*

The conditions for the release of the sphingosine bases is the same as that described for the release of the fatty acids (16).

(i)     After extraction of the fatty acids, take the aqueous phase to dryness and dissolve in 2 ml of 1.0 M NaOH.

(ii)    Extract the sphingosine bases by three additions of 4.0 ml of diethylether and evaporate the solution to dryness. For g.l.c., prepare the trimethylsilyl derivatives of the bases as described for the carbohydrate analyses.

Separation is achieved on 3% SE-30 coated on Gas-Chrom Q, using glass columns (2 m × 4 mm i.d.) and $N_2$ at a flow rate of 15 ml min$^{-1}$. The column temperature is 195°C for bases up to C18 and 240°C for those with longer chains.

### 3.1.4 *Glycerol*

(i)     Dissolve 1 mg of glycoglycerolipid in 1.0 ml of dry diethyl ether.

(ii)    Add a solution of 3 mg of lithium aluminium hydride in 0.5 ml of diethyl ether dropwise until it no longer boils.

(iii)   Add a further aliquot of the hydride and reflux the solution for 1 h.

(iv)    Add 0.5 ml of acetic anhydride dropwise to the cooled solution to destroy excess hydride.

(v)     Add 0.5 ml 'Analar' xylene and reflux for 6 h.

(vi)    Evaporate the solution to dryness and analyse by g.l.c. on an SE-30 column similar to that used for the separation of sphingosine bases.

Glycerol can also be estimated by a spectrophotometric method using a reagent mixture containing the following (all final concentrations); 71.4 mM triethanolamine buffer, 8.5 mM $Mg^{2+}$, 1.0 mM phosphoenol pyruvate, 0.4 mM NADH, 2.0 mM ATP, 20 $\mu$g ml$^{-1}$ pyruvate kinase ($>3$ U ml$^{-1}$) and 10 $\mu$g ml$^{-1}$ lactate dehydrogenase ($>3.6$ U ml$^{-1}$).

(i)     Add 0.1 ml of a solution of glycerol, obtained either by the action of lipase or alkaline hydrolysis of the glyceride, to 0.4 ml of 0.1 M triethanolamine buffer (pH 7.6) and 0.5 ml of the reagent mixture.

(ii)    Follow the absorbance change at 340 nm for 10 min.

(iii)   Add 5 $\mu$l of 5 $\mu$g ml$^{-1}$ glycerokinase ($>3.6$ U ml$^{-1}$) and record the new absorbance after a further 10 min. The difference between the two values is proportional to the glycerol content.

The coefficient of variation for the method is $2-3\%$ and only dihydroxyacetone and L-glyceraldehyde react in the same way as glycerol (17).

## 3.2 Methylation analysis of oligosaccharide chains

### 3.2.1 *Methylation procedure*

The methylation of oligosaccharide chains in intact glycolipids or those released from glycolipids has only been practical as a method of sequence analysis since the procedure of Hakomori (18) became available. Prior to 1964, methylation procedures either required too large a sample or caused too much degradation to give useful results.

The method requires the preparation of methylsulphinyl anion (see also Chapter 3, Section 6 and Chapter 5, Section 7).

(i)     Prepare dry dimethyl sulphoxide by refluxing 100 ml of commercial grade material over 2 g of calcium hydride.

(ii)    Distill at reduced pressure.

(iii)   Store the solvent over molecular sieve pellets (Linde Type 4A).

(iv)    Wash sodium hydride (1.5 g of a dispersion in mineral oil) four times either with dry pentane or petroleum ether (BP $40-60°C$) in a three necked flask and remove the oil by decantation.

(v)     Fit the flask with a condenser and thermometer and remove residual solvent by repeated evacuations using $N_2$ as the flushing gas.

(vi)    Flush the flask continually with $N_2$ while adding 15 ml of dry dimethyl sulphoxide.

(vii)   Heat the contents to 50°C with constant stirring using a magnetic stirrer for about 45 min.

The reagent is stable if stored at $-20°C$ and can be done so in 1.0 ml batches in vials under $N_2$ or argon.

(i)     Dissolve 1.0 mg of glycolipid in 1.0 ml of dry dimethyl sulphoxide in a screw-capped vial fitted with a Teflon-coated liner and containing a small magnetic stirrer bar.

(ii)    Add 1.0 ml of methylsulphinyl anion to the solution and flush the vial with $N_2$ before capping.

(iii)   Stir the mixture for at least 3 h at room temperature before adding 1.0 ml of redistilled methyl iodide. Continue stirring for a further 2 h to effect complete methylation.

(iv)   Add 1.0 ml of chloroform-methanol (1:1 v/v) and apply the sample to a Sephadex LH20 column (30 × 1 cm) equilibrated with chloroform-methanol (1:1 v/v).

(v)   Using the same solvent for elution, collect small fractions (1 −2 ml) on a fraction collector.

(vi)   Monitor the eluate for the presence of carbohydrate either by a colorimetric method such as the phenol-sulphuric acid procedure (see Chapter 1, Section 2.1.2). Alternatively spot aliquots on a t.l.c. plate and char.

(vii)   Bulk fractions containing carbohydrates together and evaporate in a stream of $N_2$.

Procedures giving improved yields of methylated products and cleaner products have been reported (19) which use potassium methylsulphinyl carbanion (prepare either from potassium hydride or potassium *t*-butoxide) instead of sodium methylsulphinyl carbanion.

Extreme caution must be taken in preparing these carbanions, particularly the potassium one. Sodium and potassium hydrides are relatively stable as dispersions in mineral oil but the pure hydrides especially potassium hydride are extremely reactive and must be used only in small quantities and with great care. The carbanions are also reactive with oxygen, carbon dioxide and water so must be stored under $N_2$ or argon.

Due to the small sample size being used, it is not practical to determine the methoxyl content by a chemical method. The most useful criterion for complete methylation is the absence of a hydroxyl peak near 3500 cm$^{-1}$ in the i.r. spectrum.

### 3.2.2 *Separation and characterization of methylated sugars*

*Hydrolysis.* Methylated glycolipids are not normally water-soluble so special hydrolytic procedures are required. The method generally acceptable, especially when hexosamines are present, is as follows.

(i)   Dissolve the methylated glycolipid in 0.3 ml of glacial acetic acid:5 M sulphuric acid (19:1 v/v) and heat at 80°C for 18 h.

(ii)   Add a further 0.3 ml of water and continue heating for a further 5 h at 80°C (20).

(iii)   Pass the hydrolysate down a small column of Dowex 50W-X8 cation exchange resin (200 mg) in the acetate form contained in a Pasteur pipette and wash the column with 4.0 ml of methanol. This removes the sulphate ion from the hydrolysate.

(iv)   Collect the washings and evaporate to dryness either in a stream of $N_2$ or on a rotary evaporator.

When only neutral sugars are present, the glycolipid can be depolymerised by heating in 1.0 ml of 95% formic acid at 100°C for 2 h. Residual formic acid is removed by distillation and the formyl esters are finally hydrolysed with 0.1 M sulphuric acid at 80°C for 18 h. The clean-up method is the same as used for the acetic/sulphuric acid method.

Another convenient method for neutral sugar chains is to autoclave with 0.5 ml of 2 M trifluoroacetic acid for 1 h at 121°C. The hydrolysate is simply evaporated to dryness in a stream of $N_2$ as the trifluoroacetic acid is volatile.

Methanolysis with hydrochloric acid in methanol can be used but there is evidence that demethylation can occur under these conditions.

*Derivatization.* The partially methylated sugars in hydrolysates are not sufficiently volatile to be analysed satisfactorily by g.l.c. In addition, direct esterification of the free sugars produces the anomeric forms of both pyranose and furanose ring forms which results in very complex chromatograms. Most workers in the field use reduction to the sugar alditol followed by acetylation to prepare volatile derivatives for subsequent analysis. The methods for preparing peracetylated partially methylated alditols is fully described in Chapter 1, Section 6.2.

The only sugar frequently found in glycolipids which can be misinterpreted by this method is D-galactose. Reduction to galactitol introduces a plane of symmetry between C-3 and C-4. The problem can be overcome by reducing the sugars with sodium boro-deuteride which allows definitive identification if mass spectrometry is used. Alternatively the methylated sugars can be converted to the fully acetylated partially methylated aldonitriles (21) but the data on the separation of these derivatives is not so complete as for the alditols.

*Analysis by g.l.c. and g.l.c.-m.s.* There are no specific problems that are unique to the analysis of glycolipids and the reader is directed to Chapter 1 for full details.

*Analysis by h.p.l.c. of peralkylated partially methylated oligosaccharides.* Complex oligosaccharide chains from glycolipids, after methylation, can be partially hydrolysed either by non-specific acidic procedures or by highly specific procedures should a particularly labile bond be present in the molecule. The released oligosaccharides can be reduced and peralkyl (usually perethyl)ated as previously described. Separation of the peralkyl partially methylated oligosaccharide alditol can be achieved by h.p.l.c. using a C-18 reverse phase column with a linear gradient of acetonitrile:water (1:1 v/v) to acetonitrile:water (13:7 v/v) over 45 min as eluent. The eluent is detected with a refractive index monitor and only a few mg of sample are required. The products can be isolated and identified further by g.l.c.-m.s. or by m.s. alone.

## 3.3 Isolation of oligosaccharide chains

### 3.3.1 *Trifluoroacetolysis*

When glycolipids are reacted with a mixture of trifluoroacetic acid and trifluoroacetic anhydride, a number of reactions take place (22). O-Trifluoroacetylation is complete and amino functions become N-trifluoroacetylated. N-Acetyl groups undergo trans-amidation also yielding N-trifluoroacetyl groups. Sialic acid residues are hydrolysed from gangliosides and ceramides which are also degraded releasing the free oligosaccharides. The yield of the oligosaccharides depends on the ratio of the acid to the anhydride. For example a 1:1 mixture gave a better yield of D-glucose and D-galactose from their respective ceramides than a 1:50 mixture while a better yield of the oligosaccharide

$$\text{D-GalNH}_2\text{-}\beta\text{-}(1 \rightarrow 4)\text{-D-Gal-}\beta\text{-}(1 \rightarrow 4)\text{-D-Glc-OL}$$

was obtained with the 1:50 mixture.

The oligosaccharides can be analysed by methylation procedures as previously described. The *N*-trifluoroacetyl group is retained during methylation and improves the separation of amino sugars on g.l.c. The procedure only applies to glycolipids which contain an unsaturated sphingosine base (23), fortunately the common occurrence. Those containing a saturated sphingosine base are not ruptured between the carbohydrate and the ceramide portion: they only undergo *O*- and *N*-trifluoroacetylation.

### 3.3.2 *Osmium tetroxide — periodic acid oxidation*

This procedure also requires the presence of an unsaturated sphingosine base.

(i)   Acetylate 1−3 mg of glycolipid with 1.0 ml acetic anhydride and 0.5 ml of pyridine at 37°C for 18 h.

(ii)  Repeatedly evaporate with toluene.

(iii) Dissolve the acetylated glycolipid in 0.5 ml of methanol and 0.2 ml of 0.2 M periodic acid in methanol.

(iv)  Add 25 $\mu$l of 5% osmium tetroxide in ether and leave overnight at 5°C.

(iv)  Add a drop of glycerol to destroy any excess reagent and partition the mixture by adding 6 ml chloroform:methanol (2:1 v/v) and 1.5 ml water.

(v)   Wash the organic phase with chloroform:methanol:water (1:10:10 v/v/v) and evaporate to dryness.

(vi)  Release the oligosaccharides from the aldehyde by dissolving in 0.3 ml of methanol and adding 50 $\mu$l of 0.2 M sodium methoxide in methanol before leaving for 1 h at room temperature.

(vii) Add a drop of acetic acid, evaporate the mixture to dryness and take up in chloroform:methanol:water (1:10:10 v/v/v) before washing with chloroform:methanol:water (60:35:8 v/v/v). The aqueous layer contains the pure oligosaccharide. Reported yields are from 44% for ceramide dihexoside to 70% for ceramide trihexoside (24).

### 3.3.3 *Treatment with ozone*

(i)   Dissolve 5 mg of glycolipid in 2.0 ml of methanol and saturate the solution slowly with ozone (prepared from an ozone generator) at 20°C (25).

(ii)  Evaporate the solvent and dissolve the residue in 2 ml of 0.1 M sodium hydroxide or 0.1 M sodium carbonate. Leave at room temperature for 18 h.

(iii) Treat the product with Dowex 50W-X8 cation exchange resin ($H^+$ form) to remove $Na^+$ ions and extract with chloroform to remove fatty aldehydes.

(iv)  Fractionate the products, in 0.1 M acetic acid, on a Sephadex G25 (fine) column (77 × 1.4 cm) with 0.1 M acetic acid as eluant.

### 3.3.4 *Determination of anomeric configuration by oxidation with chromium oxide*

This method is one of the few chemical methods which has shown much success in other branches of carbohydrate chemistry. For reasons yet unknown, it has not proved to be generally useful in glycolipid methodology since some $\beta$-glycosides are resistant to oxidation while some $\alpha$-glycosides are oxidized.

## 4. PHYSICAL METHODS OF ANALYSIS

### 4.1 **N.m.r. spectroscopy**

#### 4.1.1 *Proton magnetic resonance*

Proton magnetic resonance spectra can be obtained from native glycolipids, the oligosaccharides released from glycolipids or derivatized (such as permethylated or peracetylated) glycolipids. There are advantages in using each method but when all possible information is being sought, particularly details of the secondary structure, the native glycolipid is the sample of choice. All the $^1$H-n.m.r. signals of the ceramide component can be assigned and the oligosaccharide component neither has any effect on these signals nor gives signals which overlap with those from the ceramide component.

As native glycolipids are not generally soluble in water or chloroform, it is necessary to use other solvents. Fully deuterated dimethyl sulphoxide containing 2% $D_2O$ is the solvent of choice but some workers use pyridine-$d_5$. Unambiguous results have been obtained with as little as 100 $\mu$g of glycolipid using high resolution instruments and signal averaging techniques. Less sophisticated instruments require larger-sized samples and it is not considered likely that future generations of instruments will be able to significantly improve this lower limit. There is no hope of $^1$H-n.m.r. spectroscopy improving detection limits by factors of $10^1 - 10^2$.

The literature on $^1$H-n.m.r. spectroscopy of glycolipids is too great to be considered in full and the readers are directed to a recent review (26) for more details. Some general points are summarised below.

The H-1 signals in the $^1$H-n.m.r. spectrum are well separated from the other ring proton signals and show characteristic chemical shifts. The numbers of each type of sugar residue is determined by the intensity of the signal. Some ambiguities, such as those concerning the $\alpha$-galacto-series, can be clarified by examining other resonances of each particular member. Hence, $\alpha$-N-acetylgalactosamine gives signals from the methyl component of the acetamido group and fucose gives signals from its C-6 methyl group. Other ambiguities, such as $\beta$-glucose and $\beta$-galactose can be resolved by considering other resonance, in this example the H-2 resonances.

In addition to determining the sugar residues in a glycolipid, the H-1 signals also give the anomeric configuration of the sugar. The coupling constants $J_{1,2}$ are characteristic with values of $3-4$ Hz obtained from $\alpha$-anomers and $7-9$ Hz from $\beta$-anomers. The conformation of the sugar rings was obtained by detailed examination of monogalacto- and monoglucosylceramide.

A comparison of the other ring proton signals from an unknown glycolipid with the corresponding signals from other members of the same family of glycolipids is able to predict the sites of glycosidic linkage and the sequence of the oligosaccharide chain. Such procedures all require the use of Fourier-transform spectra coupled with spin decoupling difference spectra and spectra obtained using the nuclear Overhauser effect. An example of the assignment of the sugar residues and anomeric configuration of a series of glycolipids is shown in *Table 1*.

#### 4.1.2 $^{13}$C-n.m.r. *spectroscopy*

This technique cannot yet be used on unknown glycolipids since the amount of sample required is large relative to other methods. Most spectra have been determined on

**Table 1.** H-1 chemical shifts (p.p.m. from Me$_4$Si) and $^3J_{1,2}$ coupling constants (Hz) of glycosphingolipid in dimethyl sulphoxide-d$_6$[a].

| Compound | GalNAc (α 1-3) δ | J | GalNAc (β 1-3) δ | J | Gal (α 1-4) δ | J | Gal (β 1-4) δ | J | Glc (β 1-1) Cer δ | J |
|---|---|---|---|---|---|---|---|---|---|---|
| 1 | – | – | – | – | – | – | – | – | 4.10 | 7.7 |
| 2 | – | – | – | – | – | – | 4.22 | 7.3 | 4.17 | 7.7 |
| 3 | – | – | – | – | – | 4.0 | 4.27 | 7.7 | 4.16 | 8.1 |
| 4 | – | – | 4.52 | 8.1 | 4.81 | 3.6 | 4.26 | 7.7 | 4.16 | 7.7 |
| 5 | 4.74 | 3.6 | 4.56 | 8.5 | 4.82 | 3.6 | 4.27 | 7.7 | 4.17 | 7.7 |

[a]Ref. 34, reprinted with permission.

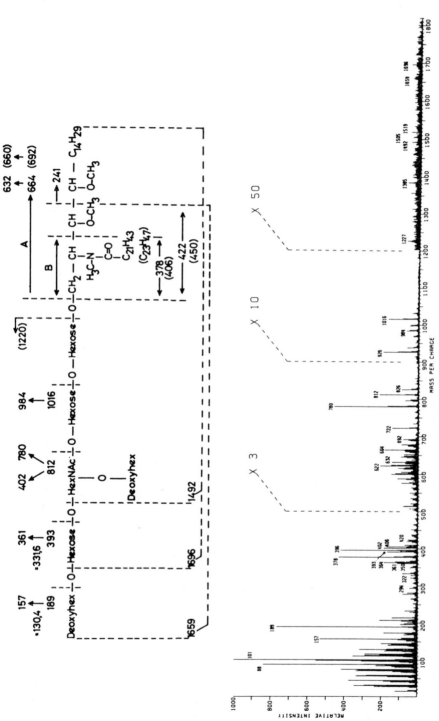

**Figure 1.** Fragmentation scheme and mass spectrum of permethylated Le$^b$-active glycosphingolipid (fraction I) from the plasma of Le(a$^-$b$^+$) donors (27). Reprinted with permission.

glycolipids whose structure is already well documented. The main advantage is that information on the conformation of individual monosaccharide residues is obtained with greater ease. Information on the identity of the sugar residues, anomeric configuration and position of linkage can all be obtained.

## 4.2 Mass spectrometry

The use of mass spectrometry for the identification of methylated sugars obtained by hydrolysis of methylated glycolipids is discussed in Chapter 1, Section 8. In this section, application of mass spectrometry to native glycolipids and oligosaccharides derived from them will be discussed.

### 4.2.1 Electron impact mass spectrometry

The permethylated glycolipid is purified by high-performance t.l.c. and 300 μg is dissolved in 40 μl of dichloromethane before placing in quartz tubes of the direct inlet probe of the mass spectrometer. The structure of a permethylated Le$^b$-active glycosphingolipid was determined by this method, basing the results on the spectrum of neolactotetraosyl ceramide whose structure had already been established by similar methods (27). The largest ion that could be positively identified had an m/e value of 1696 although many others with values >1000 were also identified (see *Figure 1*).

In principal there is no reason why the permethylated derivative must be used. It is considerably easier to prepare peracetyl, pertrimethylsilyl or pertrifluoroacetyl derivatives. The practical limitations are the resolving power and range of the mass spectrometer. Since each methyl group only adds 14 mass units to the molecule while an acetyl group adds 42 units and the others more, permethylation allows the analysis of glycolipids of considerably longer chain length.

### 4.2.2 Fast atom bombardment mass spectrometry

This relatively new technique will probably be the method of choice for structural analysis when it becomes more available. The sample size is only about 5 μg which is dissolved in a suitable solvent (0.5 μl) and added to a drop of glycerol on the target. The target is bombarded with Argon or Xenon atoms of 8 KeV and the spectrum recorded. The mass spectrum of an unmodified tetradecasaccharide gave a molecular ion [(M + Na)$^+$] of m/e 2555 (28). In another investigation, negative ion fast atom bombardment m.s. of native glycosphingolipids and positive ion fast atom bombardment m.s. of permethyl and peracetyl glycolipids have complemented $^1$H-n.m.r. results to determine the molecular weight, ceramide moiety, type and number of sugar constituents, their anomeric configuration, sequence, sites of linkages and patterns of branching in complex glycosphingolipids (29). See also Chaper 5, Section 7.4.

## 5. BIOCHEMICAL METHODS OF ANALYSIS

### 5.1 Enzymatic methods

It should be possible to determine the sequence and anomeric configuration of oligosaccharide chains of glycolipids by sequential degradation with specific exo-glycosidases and some structures have been proposed by this method and later

GalNAc-α-(1→3)Gal-β-(1→3)-GlcNAc-β-(1→3)-Gal-β-(1→4)-Gal-β-(1→4)-Glc-β-(1→1)-Cer
        2
        ↑
        1
     α-Fuc

**Figure 2.** Structure of a glycolipid determined by sequential enzymatic hydrolysis (30).

corroborated by other techniques. Only about 100 μg of glycolipid is required but the reaction mixture buffer (actual conditions depend on the enzyme being used) requires to be supplemented with for example sodium taurocholate to keep the glycolipid in solution. The sequential use of α-*N*-acetylgalactosaminidase, β-*N*-acetylhexosaminidase, α-L-fucosidase, β-D-galactosidase and β-*N*-acetylglucosaminidase on a native A-active glycolipid from hog stomach mucosa has shown that it has the structure in *Figure 2* (30). In many examples, the sugar residue released by the enzyme can be quantitatively recovered by partitioning between an organic and an aqueous phase. The glycolipid containing one fewer sugar residue remains in the organic phase and can easily be recovered and treated with another enzyme.

There are reported failures due to varied specificities of the enzymes used. For example the sialic acid residue of $GM_1$ ganglioside could not be hydrolysed by available sialidases but it was later found that the sialidase from *Arthrobacter ureafaciens* was able to bring about hydrolysis in the presence of non-ionic detergents (31). Other enzymes are reported to have their own specificity.

## 5.2 Immunological methods

The methodology and implications of the use of immunological techniques are too diverse to be considered in detail but attention is drawn to a recent review (32). As an example, two monoclonal antibodies were isolated by the murine hybridoma technique and their reactivity determined by conventional monoclonal antibody methods (33). One antibody was specific for non-reducing terminal *N*-acetyl lactosamine structures [Gal-β-(1→4)-GlcNAc-β-(1-R)] and the other was specific for the fucosylated *N*-acetyl lactosamine structure [Fuc-α-(1→2)-Gal-β-(1→4)-GlcNAc-β-(1-R)]. One major use of such antibodies is in investigating the distribution, quantity and function of glycolipids on the cell surface.

## 6. REFERENCES

1. Folch,J., Lees,M. and Sloane Stanley,G.H. (1957) *J. Biol. Chem.*, **226**, 497.
2. Suzuki,K. (1965) *J. Neurochem.*, **12**, 629.
3. Slomiany,B.L. and Slomiany,A. (1977) *Biochim. Biophys. Acta*, **486**, 531.
4. Tettamanti,G., Bonaldi,F., Marchesini,S. and Zambotti,V. (1973) *Biochim. Biophys. Acta*, **296**, 160.
5. Baumann,H., Nudelman,E., Watanabe,K. and Hakomori,S. (1979) *Cancer Res.*, **39**, 2637.
6. Wuthier,R.E. (1966) *J. Lipid Res.*, **7**, 558.
7. Rouser,G., Kritchevsky,G., Yamamoto,A., Simon,G., Galli,C., Bauman,A.J. (1969) in *Methods in Enzymology*, Lowenstein,J.M. (ed.), Academic Press Inc., New York and London, Vol. **4**, p. 272.
8. Yu,R.K. and Ledeen,R.W. (1972) *J. Lipid Res.*, **13**, 680.
9. Momoi,T., Ando,S. and Nagai,Y. (1976) *Biochim. Biophys. Acta*, **441**, 488.
10. Stellner,K., Watanabe,K. and Hakomori,S. (1973) *Biochemistry*, **12**, 656.
11. Hakomori,S., Nudelman,E., Levery,S. and Kannagi,R. (1984) *J. Biol. Chem.*, **259**, 4672.
12. Evans,J.E. and McCluer,R.H. (1972) *Biochim. Biophys. Acta*, **270**, 565.
13. Gross,S.K. and McCluer,R.H. (1980) *Anal. Biochem.*, **102**, 429.

14. Watanabe,K., Hakomori,S., Childs,R.A. and Feizi,T. (1979) *J. Biol. Chem.*, **254**, 3221.

15. Chambers,R.E. and Clamp,J.R. (1971) *Biochem. J.*, **125**, 1009.

16. Weiss,B., Stiller,R.L. and Jack,R.C.M. (1973) *Lipids*, **8**, 25.

17. Eggstein,M. and Kuhlmann,E. (1974) in *Methods of Enzymatic Analysis*, Bergmeyer,H.U. (ed.), Academic Press Inc., New York, Vol. **4**, p. 1825.

18. Hakomori,S. (1964) *J. Biochem.*, **55**, 205.

19. Harris,P.J., Henry,R.J., Blakeney,A.B. and Stone,B.A. (1984) *Carbohydr. Res.*, **127**, 59.

20. Stellner,K., Saito,H. and Hakomori,S. (1973) *Arch. Biochem. Biophys.*, **155**, 464.

21. Dmitriev,B.A., Backinowsky,L.V., Chizhov,O.S., Zolotarev,B.M. and Kochetkov,N.K. (1971) *Carbohydr. Res.*, **19**, 432.

22. Svensson,S. (1980) in *Mechanisms of Saccharide Polymerization and Depolymerization*, Marshall,J.J. (ed.), Academic Press Inc., New York and London, p. 387.

23. Lindh,F. and Nilsson,B. (1984) in Abstracts of XII International Carbohydrate Symposium, p. 447.

24. MacDonald,D.L., Patt,L.M. and Hakomori,S. (1980) *J. Lipid Res.*, **21**, 642.

25. Wiegandt,H. and Bucking,H.W. (1970) *Eur. J. Biochem.*, **15**, 287.

26. Dabrowski,J., Hanfland,P. and Egge,H. (1982) in *Methods in Enzymology*, Ginsburg,V. (ed.), Academic Press Inc., New York and London, Vol. **83**, p. 69.

27. Egge,H. and Hanfland,P. (1981) *Arch. Biochem. Biophys.*, **210**, 396.

28. Dell,A., Morris,H.R., Egge,H., Von Nicolai,H. and Strecker,G. (1983) *Carbohydr. Res.*, **115**, 41.

29. Egge,H., Dabrowski,J. and Hanfland,P. (1984) *Pure Appl. Chem.*, **56**, 807.

30. Slomiany,B.L., Slomiany,A. and Horowitz,M.I. (1974) *Eur. J. Biochem.*, **43**, 161.

31. Saito,M., Sugano,K. and Nagai,Y. (1979) *J. Biol. Chem.*, **254**, 7845.

32. Hakomori,S. and Young,W.W.,Jr. (1983) in *Handbook of Lipid Research. 3. Sphingolipid Biochemistry*, Kanfer,J.N. and Hakomori,S. (eds.), Plenum Press, New York and London, p. 381.

33. Young,W.W.,Jr., Portoukalian,J. and Hakomori,S. (1981) *J. Biol. Chem.*, **256**, 10967.

34. Dabrowski,J., Hanfland,P. and Egge,H. (1980) *Biochemistry*, **19**, 5652.

# INDEX